Overcoming Overtourism

Bruno S. Frey

Overcoming Overtourism

Creating Revived Originals

 Springer

Bruno S. Frey 🆔
University of Basel
Basel, Switzerland

CREMA - Center for Economics
Zurich, Switzerland

ISBN 978-3-030-63813-9   ISBN 978-3-030-63814-6   (eBook)
https://doi.org/10.1007/978-3-030-63814-6

This Springer imprint is published by the registered company Springer Nature Switzerland AG.
The registered company address is: Gewerbestrasse 11, 6330 Cham, Switzerland

Acknowledgments

Discussions with my colleagues, acquaintances, and friends helped me to come to grip with the problem of cultural overtourism, especially with my proposal of Revived Originals. Some are skeptical; others think it is a good idea. Above all, I would like to thank Martin Beglinger, Reto Cueni, Jens Drolshammer, Reiner Eichenberger, Gerd Folkers, Beat Gygi, Jürg Helbling, Alexander Hunziker, Leopold Kohn, Kai Konrad, Simon Lüchinger, Christine Maier, Peter Nobel, Jan Osterloh, Lasse Steiner, Christoph Schaltegger, Urs Schupp, Angel Serna, Alois Stutzer, Benno Torgler, Christian Ulbrich, Joachim Voth, Anne-Lea Werlen, and Oliver Zimmer.

I am also grateful for the many useful comments I received during my lectures on the subject at various universities and research centers.

This book partly follows an earlier publication in German entitled *Venedig ist überall. Vom Übertourismus zum Neuen Orginal* (Frey 2020) and builds on various articles to be published in scientific journals. One of these has partly been written together with my assistant Andre Briviba (Frey and Briviba 2020). The idea of Revived Originals is also briefly discussed in Chap. 14 on "Cultural Tourism" in the book of Frey (2019).

Simon Milligan, Evelyn Holderegger, Sonja Helfer, Mandy Fong, and Andre Briviba reviewed the manuscript carefully and helpfully and made many improvements possible; I am very grateful for that.

I am especially indebted to my brother René L. Frey and my wife Margit Osterloh for their many important suggestions and thoughts.

Contents

1

Overtourism: Problems and a Radical Proposal—An Introduction

When I was 8 years old, my parents offered me a trip to Paris because I was eager to see the Eiffel Tower. I was greatly impressed by this imposing monument. However, my parents emphasized that I should also see the most famous painting in the world. I found the Mona Lisa beautiful but was surprised at how small the masterpiece is in reality.

Some time ago, I wanted to see the Mona Lisa again, because I had read a book on Leonardo da Vinci and wanted to look once more at the original painting. But the situation was very different from my first visit in many ways.

I waited a long time in the blazing summer heat to enter the pyramid of the Louvre. This is no exception but rather the rule, as Fig. 1.1 shows. When I finally reached the vestibule of the museum, I immediately realized where the masterpiece is situated because a huge crowd of people was moving in the direction indicated by a sign saying "This way to the Mona Lisa." On the way in this direction, there was no stopping; I was literally pushed by the crowds. In front of the gallery, there was another queue until I finally arrived at my destination. The Mona Lisa is secured behind bulletproof glass and guarded by several security agents. Only 10–15 people are allowed to step in front of the painting at any one time. Few visitors looked at the picture, but they all tried to take a selfie with the Mona Lisa. After a few seconds, they were asked to move on, and no objection or protest was accepted. Those who do not move quickly enough are pushed away by the supervising staff. This haste is necessary because 80% of the up to 30,000 daily visitors to the Louvre want to see the Mona Lisa at all costs; the other masterpieces in the museum are apparently unimportant for most of them.

B. S. Frey, *Overcoming Overtourism*, https://doi.org/10.1007/978-3-030-63814-6_1

Fig. 1.1   Queue in front of the Louvre Museum in Paris. Source: Picture Drago Gazdik on Pixabay

The hype around the Mona Lisa is not an isolated case in cultural tourism. A friend told me that while on a trip through the Bavarian region of Allgäu, he wanted to show his wife Neuschwanstein Castle. But he soon had to turn back because all the parking lots were completely full. This is not surprising, as 1.5 million people per year want to visit this castle, which was only partially completed at the end of the nineteenth century and thus is far from being historic. On some days, more than 10,000 tourists visit King Ludwig II's "fairytale castle." It is the most prominent tourist attraction in Bavaria and far beyond. Due to the high volume of visitors, guests without an online reservation have to endure a wait of several hours. In the summer months especially, the traffic situation around Neuschwanstein Castle is extremely tense. The desperate search for parking spaces is a nuisance for residents, and the congested traffic in the nearby city of Füssen is due mainly to the arrival and departure of castle tourists.

The best-known example of cultural overtourism is Venice, which on some days is flooded by more than 120,000 visitors. About 30 million people visit the city yearly; back in 1949, the figure was less than half a million. This massive influx of tourists leads to long queues in front of the major attractions. Figure 1.2 shows the queue in front of St. Mark's Dome. In recent years, the number of residents has fallen to around 50,000 people, while the number of second homes has risen sharply. The influx of visitors includes approximately 40,000 passengers from up to four huge cruise ships that pass every day

Fig. 1.2 Overtourism in Venice; here in front of St. Mark's Dome. Source: Picture by M W on Pixabay

through the narrow channel between Venice and the island of Giudecca to dock in the port.

Overtourism also affects smaller towns. On some days, for example, more than 10,000 tourists crowd into the tiny village of Hallstatt in Austria, which has only 750 residents.

Even in a big city such as Rome, overtourism changes things noticeably. There is often a long queue in front of the Vatican museums. Most of the up to 30,000 visitors want to see the Sistine Chapel (*Capella Sistina*) with the famous frescoes by Michelangelo. Accordingly, people are squeezed in like sardines. It is not even possible to admire the wonderful paintings by Raphael in the four *Stanze*, or large rooms, of the Vatican Palace. Instead, the huge stream of people pushes one towards the *Sistina*. Once there, visitors are urged by the supervisory staff to leave as quickly as possible. Even in front of St. Peter's Basilica, there is a long queue. Surprisingly, even priests, monks, and nuns have to queue like everybody else although they are part of the organization of the Catholic Church.

Economics offers a ready solution to deal with any situation in which, as with overtourism, demand clearly exceeds supply. The visiting costs have to be increased in order to reduce the demand for the limited space available at cultural sites. In fact, various attempts have been made to charge entrance fees for visiting overcrowded cities. However, this method is rarely applied. It is only possible for places with few, and easily controllable, entrances. This is the case for Venice, where entry is only possible over one bridge or by sea; Dubrovnik, which is surrounded by a medieval wall with only five gates; and Mont Saint Michel, an island that is only accessible over a causeway.

Admission prices are often proposed and politically discussed for such cities. Nevertheless, lawyers and members of the public administration prefer to limit the number of visitors directly, or at least to limit their length of stay. This obviously causes many problems, too. Administrative decisions have to be made about who is allowed into a city. For example, are people allowed to visit friends or even stay with them overnight? An arbitrary threshold has to be defined. Moreover, people who are denied access to a cultural site and wish to visit it are deprived of these experiences; this undesirable effect must also be taken into account.

In this book, I present a radically different solution to cultural overtourism: to *increase the supply* of cultural sites rather than to restrict demand. I propose that the most important sites should be replicated identically in new locations and their attractiveness increased by means of digital technologies, in particular augmented and virtual reality as well as holograms. In this way, the history and culture of the site can be conveyed in a manner that is both exciting and instructive. At the same time, a suitable range of restaurants, hotels, and souvenir shops can be provided, and arrival and access can be designed to be more environmentally friendly and efficient.

At first sight, this scenario may seem to be absurd. One might be reminded of Disneyland or other theme parks. For some well-educated people, such an idea may even verge on the blasphemous. However, visitors to a historically replicated site would be given a unique sense of what that historical site looked like in the past and how the population lived at that time. In this respect, a replica would offer visitors more than the original site. At the same time, people could see how attractively art can be displayed in the digital world.

I will show that the proposal, which I have termed *Revived Originals*, has many advantages and is a useful way of dealing with the masses of tourists already existing and expected in the future. To date, this proposal has not been found in the literature on overtourism; expanding the supply of a cultural attraction is generally thought to be impossible. Although similar ideas have been mooted, these have to date only been intended to preserve specific cultural goods rather than whole cities. There are many negative external effects of overtourism that are not taken into account by simply restricting demand.

I am confident that many tourists will gladly accept the replica of a cultural site. The Revived Originals are not designed to be substitutes of the original sites, not least because the replicas additionally create a closer relationship with history and culture. This new offer reduces the pressure on the originals and thus makes their visit more pleasant for art lovers. One should also bear in mind that the difference between an "original" and a "copy" has long been disputed, particularly in the history of art. It a matter of dispute, for instance, whether the antique tower in the Venetian St. Mark's Square, rebuilt in 1902 after it collapsed, is an original or a copy.

Could a Revived Original encourage its visitors to see the original later? This might also fuel cultural overtourism additionally. Some tourists would certainly want to see the original after seeing the Revived Original. This is a positive effect: a new category of people will be attracted to culture.

However, it is not to be expected that the originals will be confronted anymore with overtourism once the Revived Originals are in place. Several reasons can be mentioned: Many tourists from other continents will visit Europe only once and therefore will never see the original places. Moreover, many visitors of a Revived Original will not wish to visit the original. They will be satisfied with the replica, not least because it offers more amenities and is more engaging. Rather, they will visit other of the countless cultural sites existing in the world. A reverse effect is even possible. Anyone who has seen the original may want to visit the Revived Original with its advanced digital technologies. People who like the replica better than the original may also visit Revived Originals of other cultural sites in the future. The original sites will then be less damaged by overtourism.

Revived Originals are also ecologically advantageous, as the new cultural sites are designed and operated in a particularly environmentally friendly manner. Today's overcrowded cultural sites, by contrast, are threatened by wear and tear, mutilation, and pollution. For example, the air in Venice is extremely polluted even though the city benefits from the sea breeze. The reason for this is the many cruise ships and their enormous engines, which run even while the ships are docked.

In addition, cultural tourists are obliged to visit many different places to enjoy the most important cultural sites in, for example, northern Italy. A historical replica of the most important sites of, for example, Verona, Siena, Pisa, Padua, Bergamo, and Vicenza in one place, as proposed in this book, would reduce the ecological impact of the visitors considerably.

A book about overtourism may come as a surprise, especially at this time. In winter 2019/2020, the COVID-19 pandemic started, putting an abrupt and unforeseeable end to overtourism. As this book goes to press, many countries have blocked international tourism, and in some nations even travel within their own borders has been banned. The result is the exact opposite of overtourism. Tourism has come to a standstill. Consequently new pictures have appeared in the media: a completely empty St. Mark's Square in Venice and a depopulated Great Wall in China.

The collapse of international and local tourism to cultural sites, which may be called "cultural undertourism," was fueled by the closure of museums, galleries, theaters, art festivals, and other artistic activities. This collapse gives us the chance to reconsider how visits to artistic sites should be organized in the future. Normally, traditional thinking and interest group opposition block major strategic changes. The drastic shock imposed by the COVID-19 and the resulting shutdown of the economy enables us to discuss new ventures and to embark in new directions. Moreover, as the recovery of mass tourism can be expected to take substantial time, it is possible to extensively consider and evaluate new ideas.

Thus, the appearance of this book is particularly timely, for three reasons.

Firstly, it is to be expected that international and national tourism will recover once the pandemic is over. However, it is uncertain how long it will take before situations of cultural overtourism reappear. A few experts expect a rapid recovery of tourism, while others believe it will take several years before the flow of tourists returns to levels similar to those before 2020.

Secondly, the break will allow us to think more deeply about future cultural tourism. It gives us time to consider how to avoid the deleterious effects of overtourism. The period of undertourism demonstrates that smaller numbers of tourists make many cultural sites more attractive. The suggestion made here of expanding supply by means of Revived Originals is designed to contribute to this endeavor.

Thirdly, the effects of COVID-19 have led to increased use of digital techniques. Although this trend was already noticeable before, it was intensified when people were isolated in their homes. For example, opera houses have offered free virtual performances. The Zurich Opera House put Vincenzo Bellini's *I Capuleti e i Montecchi,* Giuseppe Verdi's *Nabucco* and *Rigoletto*, and Alban Berg's *Wozzeck* on the internet. Streaming operas and ballets has certainly brought many people into contact with these great works who would probably never have gone to the opera house to see them. These broadcasts on the internet have also opened up new qualities and dimensions. On the one hand, all the audience has an excellent view on the stage, so that a particularly successful production is shown to its best advantage and is duly noticed. In addition, the digital transmission allows close-ups of singers that are impossible to see in a conventional opera performance. Those who not only sing well but also have outstanding acting skills increase the quality that an artistic performance conveys. Many viewers will appreciate these advantages of digital transmissions. While some will be unlikely to go to the opera in the future, they are still connected to the cultural experience.

In activities that were previously only possible in real life, the internet allows us to go one step beyond. One example is the organization and implementation of virtual football games and car races. Many people are involved and engaged in this new trend, not only computer nerds. They buy the corresponding computerized machines. Therefore, it has already become profitable to commercialize such activities, which makes them more real. The adaptation of this trend can be imitated by the cultural sector, which is already partly taking place.

The Revived Originals I propose in this book create new real and virtual cultural sites and transform a visit into a unique and as yet unknown intellectual and emotional experience. The idea goes far beyond anything that exists today.

Part I

Overtourism: The Problem

2

Excessive Cultural Tourism

2.1 What Are the Figures?

According to the United Nations World Tourist Organization, in 1950 (UNWTO 2018), there were 25 million international tourists and in 2018, 1.4 billion: 56 times as many. In the coming years, a further increase in international tourism is expected. However, it should be noted that forecasts were made before the major crisis in tourism imposed by the policy measure due to COVID-19.

Nevertheless, there is a widespread consensus that mass tourism will grow further in the future. Following a cautious forecast, the number of visitors to Europe will grow further from somewhat above 500 million people in 2010 and is expected to reach about 850 million in 2030 (UNWTO 2018). This means a 60% increase in only 20 years (Kester 2016; Croce 2018). The tourist sector is worldwide one of the most rapidly growing sectors of the economy. It contributes significantly to employment, infrastructure, and export revenue (Becker 2013).

In addition to the countries where tourists mainly come from (the United Kingdom, Germany, the United States, and Japan), rapidly climbing numbers of tourists are expected from China, India, oil-producing Arab countries, Indonesia, South Korea, Vietnam, Russia, and Latin America. In China, only 7% of the population has a passport, but in numbers, this equals almost 100 million people (Becker 2013). It is widely expected that the Chinese middle class will mushroom in the coming years so that hundreds of millions of people will be able to travel internationally. Some destinations are already overcrowded by tourists, such as Lisbon with 11 million visitors per year,

© The Author(s), under exclusive license to Springer Nature Switzerland AG 2021
B. S. Frey, *Overcoming Overtourism*, https://doi.org/10.1007/978-3-030-63814-6_2

Stockholm and Dublin with almost 10 million, and Copenhagen with more than 8 million, Amsterdam receives almost 16 million day visitors per year. The result is illustrated in Fig. 2.1. The city administration stopped advertising for the town so as not to attract even more tourists.

The average length of stay of international travelers has been decreasing. In most countries, the time that visitors spend at the same place has decreased by 15% in the 20 years since 1995 (Gössling et al. 2018). So the number of tourists increases significantly, but they remain at the same places for shorter periods. This empirically well-documented change (Peeters et al. 2018: 29) results in a different way of visiting cultural sites than was customary in former times, when tourists often stayed for weeks in one place. Today, visitors focus on the most important attractions. Some Chinese tourists visit six European countries within 5 days. These effects are boosted by social media, which show rankings of places to visit. They identify what the "most important" or "the best" sites are, and as long as an internet connection is available they are accessible anywhere, anytime (Garcia-Palomares et al. 2015; Ram and Hall 2017).

Since the summer of 2017, this huge increase in the number of tourists has been called "overtourism"—a rather negative expression. The term has also been widely adopted in the scholarly literature (Martín Martín et al. 2018; Koens et al. 2018; Seraphin et al. 2019; Pechlaner et al. 2019; Jacobsen et al.

Fig. 2.1 Masses of tourists in Amsterdam. Source: Picture by Dimitri Houtteman on Pixabay

2019). Since the breakdown of tourism in 2020, one may also speak of a period of "undertourism."

It is impossible to provide a clear-cut definition of overtourism because many different aspects, and many different groups are involved (e.g., Weber 2017; Weber et al. 2017). Excessive tourism affects the welfare of the local population negatively because visitors overuse the infrastructure and damage both the natural environment and cultural heritage. The excess number of tourists also stresses other tourists, who are unable to visit famous sites because of long queues and who miss the particular local atmosphere they are seeking.

Overtourism can therefore be understood to be the consequence of tourism that exceeds the physical, ecological, social, and economic capacities of a particular time and location (Jin et al. 2016; Peeters et al. 2018: 22). Wherever those characteristics apply, tourism is too intensive. However, the question remains how overcrowded a place must be to be called "overtouristed." Until recently, the literature (Throsby 2001) made a distinction between mass tourism, meaning a huge number of people, and cultural tourism with relatively few people. Today, this distinction no longer applies because a very large number of people participate in cultural tourism.

This dynamic growth in cultural tourism has been fueled by the increasing level of education, globalization, easier access to information, and rising interest in foreign cultures and heritage (Du Cros and McKercher 2020). Moreover, a new categorization of cultural activities, which previously were part of general sightseeing, has supported this growth. In the 1990s, cultural tourism was perceived as beneficial and was considered to conserve culture and to stimulate the economy (Richards 2018). Cultural tourism evolved from a focus of the educational elite towards a mass market. By now, cultural tourism exhibits other typical characteristics of mass tourism, such as high tourist numbers, international tourism operators' market power, and the burden on the local sociocultural environment. Nonetheless, various niches have been created in cultural tourism, thanks in part to its heterogeneous characteristics. These cover an enormous thematic scope, including film tourism, wine tourism, religious tourism, and art tourism (Richards 2014).

As of today, few serious analyses of cultural overtourism have been undertaken. In most cases, the literature simply refers to the phenomenon of tourists drastically overcrowding cultural sites. Overtourism can be seen as an extreme case of the tragedy of the commons (Hardin 1968; Ostrom 1990). Tourists behave as free riders that only consider the cost of their own visits. They do not take into account the (external) cost imposed on other tourists in the form of queues, stress, and disrespectful behavior, or the burden imposed on the local population by an overcrowded infrastructure, noise, pollution, crime, stress, and other disamenities. Too many tourists visit cultural sites, and other tourists and local residents are annoyed by the negative consequences.

The world map of overtourism presented by the organization *Responsible Travel and Google* (2019) lists no less than 98 destinations in 63 countries subject to overtourism. The European Union even lists 105 sites in Europe (Kamm-Sager 2019). This discrepancy shows that overtourism is difficult to pin down in part because of its diversity. Various indicators are used to quantify overtourism. The Index of Tourism Density gives an impression. In Salzburg, there are almost 400 tourists per km^2 a day, in Copenhagen 316, in Venice 158, in Bruges 140, and in Lucerne as many as 898 (Peeters et al. 2018: 86–87). However, this indicator depends strongly on the size of the city for which density is calculated. This is why Lucerne seems to be less affected than Venice, which suffers much more from the consequences of overtourism. When we speak of Lucerne, is it the quite small old town only (where tourists mostly roam), the city as an administrative unit, or the city including its suburbs? Another index shows the number of visitors per resident and is called the Index of Tourism Intensity (D'Eramo 2017). But this index also depends on which city size is considered to be relevant. Overtourism in many cases only affects some parts of a city, mostly the old town.

2.2   Cultural Overtourism

Both the existing mass tourism and the mass tourism likely to be expected in the future are closely connected with culture (Steinecke 2010; Kaminski et al. 2013; Richards 2018; UNESCO 2019). The term "cultural tourism" is well established (e.g., Richards 1996; Leslie and Sigala 2005; Smith and Richards 2013). Cultural sites are a major reason to visit other countries and attract millions of travelers each year. This interest is also reflected in the increasing number of regional music festivals and other organized cultural events.

There are certainly also tourists whose main focus is on the beauty of the natural environment as well as those indulging in sports vacations. Overtourism exists in these areas, too. Sometimes this happens in an extreme form. An example is the large number of people who want to climb Mount Everest, the highest peak of the world (8848 m). When the weather is suitable, climbers even have to queue to reach the summit. Whereas in the period 1950–1954 there was at the most some dozens, from 2015 to 2019, there were no less than 5000 climbers. They hinder each other so severely that in 2019 four of the eleven deaths on Mount Everest during the spring climbing season were blamed on overcrowding (*The Economist* 2020a).

Another striking example of this kind of tourism is skiing resorts. Considering the development of climate warming, such places located at lower altitudes will not be able to offer their services in the future. Because of the continually decreasing likelihood of snow, the concentration of tourism in skiing resorts located at higher altitudes will increase and likely show symptoms of overtourism (Steiger et al. 2019). Even if the concept of Revived Originals has not been applied to this kind of tourism, the evaluation of measures aiming to counter the negative effects of overtourism could prove useful in this setting, too.

Most tourists visiting metropolises such as Paris, London, Saint Petersburg, Madrid, Amsterdam, Vienna, and Berlin are eager to pay a visit to the world-famous museums located there (Frey and Meier 2006). They are indeed frequented by millions of people. Table 2.1 shows the numbers of visitors for 2018.

Many tourists are satisfied with visiting only one museum. But some are determined to visit two or even more. In London, this would above all be the Tate Modern, the National Gallery, the British Museum, and the Victoria and Albert Museum; in Madrid, the Reina Sofia Museum and the Prado. There are many other highly visited museums, for instance, for technology and transportation, and whole museum complexes such as the Museum Island in Berlin including the Alte Museum, the Neue Museum, the Alte Nationalgalerie, the Bode-Museum, and the Pergamon-Museum. In 1999, Berlin's Museum Island was included in the UNESCO list of world heritage sites.

In addition to museums, culturally interested tourists also enjoy presentations of classical music, for instance at the Scala in Milan, the Viennese State Opera, Covent Garden in London, and the Bolshoi Theatre in Moscow.

Table 2.1 Number of visitors to famous European art museums (2018, in millions)

Musée du Louvre	Paris	10.2
Vatican Museums	Rome	6.8
Tate Modern	London	5.9
British Museum	London	5.8
National Gallery	London	5.7
State Hermitage	St. Petersburg	4.2
Victoria and Albert	London	4.0
Reina Sofia	Madrid	3.9
Prado	Madrid	3.7
Centre Pompidou	Paris	3.6

Source: *The Art Newspaper* (2019), https://www.museus.gov.br/wp-content/uploads/2019/04/The-Art-Newspaper-Ranking-2018.pdf

Festivals of operas and classical music are another great attraction, such as the Salzburg Festivals and the Bayreuth Festival, where the operas of Richard Wagner are presented.

My analysis of the recent extent of cultural overtourism has so far referred mainly to Europe. Of course, there are also important cultural mass attractions on other continents, for instance, the pre-Columbian ruins at Machu Picchu in Peru, the pyramids of Chichén Itzá in Mexico, the temple Angkor Wat in Cambodia, the many cultural sites in Egypt, and the Taj Mahal mausoleum in India. Figure 2.2 shows the Great Wall in China, badly overcrowded. These sites are all on the list of UNESCO's World Heritage Sites and are visited by an enormous number of people. For instance, on some days, 70,000 people come to see the Taj Mahal.

In addition to large cities such as Amsterdam, Barcelona, Berlin, and Prague, cultural mass tourism also occurs in in smaller cities such as Salzburg in Austria, Granada in Spain, Strasbourg and Colmar in Alsace in France, Reykjavik in Iceland, Lucerne in Switzerland, and Dubrovnik in Croatia. (García-Hernández et al. 2017; Hospers 2019). In 2018, 3 million tourists visited Dubrovnik and 400 cruise ships docked there. This huge number is partly because parts of the fantasy series *Game of Thrones* were shot in that town (Connolly 2019).

Fig. 2.2 Tourists on the Great Wall, China. Source: Picture by StockSnap on Pixabay

In 2018, 26 million people worldwide booked cruises visiting cultural cities. In the future, even more are expected (Senn and Egger 2019). Medium-sized cities such as Florence, Siena, Padua, Pisa, and Verona in Italy are therefore full of cultural tourists.

Even small places are affected by cultural overtourism. Examples include Stratford-upon-Avon, Cambridge and Oxford in England, Rothenburg ob der Tauber in Germany, San Gimignano in Italy, Brügge in Belgium, Český Krumlov in the Czech Republic, Interlaken in Switzerland, and Hallstatt in Austria. The castles in the Loire region of France are another example of an extremely overcrowded cultural destination in Europe.

2.3   Why Is Cultural Overtourism Growing So Quickly?

The rapid increase of cultural mass tourism can be attributed to seven major causes (Dodds and Butler 2019; UNWTO 2018): higher income, more leisure time, lower travel costs, more travel by cruise ships, hype started by social and traditional media, inclusion in the UNESCO World Heritage List, and popularity through film and media. These reasons are now discussed in turn.

1. *Growing Incomes in Threshold Countries*
   The development of the global economy sharply increased the income of many millions of people, allowing them to engage in cultural tourism. The rise of the middle class in China is particularly significant. Each year 20–30 million more Chinese become eligible for a passport and thus to visit foreign countries and continents (Neuhaus 2019). A similar development is to be found in India, where an increasing number of families have left poverty behind and gained the financial means to travel abroad. In the oil-producing Arab countries, many people are sufficiently well off to be able to visit cultural sites abroad. Indonesians and South Americans exhibit a similar development. It is to be expected that the African Continent can also achieve sufficient economic growth to allow more of its citizens to enjoy cultural tourism.
2. *Increasing Leisure Time*
   With increasing income, working time tends to be reduced. People then have more leisure time available to visit cultural sites abroad. In the past, most of these journeys were undertaken in groups, partly organized by employers. But there are signs that in the future a growing percentage of

journeys will be undertaken individually. Language barriers today can be overcome quite easily with language apps. Thus, with the help of smartphones, menus written in German and other languages can be translated into Chinese within seconds.

3. *Falling Travel Costs*

International travel has become much cheaper. More intercontinental flights are also available at low fares. Airbnb has reduced the expenses for overnight accommodation. This unregulated or at least less regulated, form of overnight stay has been growing rapidly in recent years. In Florence, Italy, 21.4% of the apartments in the historic center were offered in 2018 on Airbnb (Phelan 2018).

4. *Cruise as a Mass Phenomenon*

Ever more people book cruises traveling to the most important cultural sites. In Venice and Dubrovnik, often three to four colossal cruising ships dock at the same time, each enabling 5000 passengers or more to visit the cultural centers on the same time. In 2018, 1.6 million passengers on 466 cruising ships visited Venice (Association of Mediterranean Cruise Ports 2018). In the future, cruise ships with a higher capacity of up to 9500 passengers may be built and will take tourists to cultural sites.

5. *Selfie Culture in Social Media*

Social media create hypes for certain cultural sites. They must be "instagrammable," meaning they must look attractive. Many tourists find it important to share their selfies with friends and colleagues, to inform them that, for instance, they are standing in front of the Mona Lisa (Zeng and Gerritsen 2014; D'Eramo 2017). Selfies also provide a sort of immortality (Reinhardt 2019). "Geo-tagging" links the site of a picture via smartphone with the geographical coordinates and automatically marks them. In 2017, according to a survey of millennials, 40% indicated that they wish to communicate the choice of their travel destination on Instagram. More than 60% want to share their travel experiences on social media. There is already an app that searches for convenient flights to the Instagram pictures with one click. Travel destinations and the corresponding pictures that are shared on social media are visited much more often than before. For example, some years ago, 800 enthusiasts per year visited Trolltunga, a prominent rock in the West of Norway. Recently, 15,000 people "liked" a picture of an influencer on Instagram, which made the place world famous and resulted in 80,000 people traveling there to see it (Pousset 2019).

This signaling effect is particularly relevant for Chinese tourists. Traveling to Venice, Rome, or Florence is an important experience in their lives. Cultural and natural sites known through social media are visited, and it is essential to show this to friends and colleagues at home. Shooting and posting a picture of the sites known to them is more important than really exploring them and delving into them (Neuhaus 2019).

Traditional media also take part in this activity, for instance by regularly publishing rankings of the most frequently visited cultural sites. Attention is therefore focused on the most prominent cultural sites, further accentuating their importance in public (Goodwin 2017; Koens et al. 2018). These places attain the character of "superstars" (Rosen 1981); they even become more prominent than similar well-known cultural sites. This branding increases overtourism even more (Seraphin et al. 2018a, 2019).

6. *Particular Labels Increase Attractiveness*

Empirical studies show that sites mentioned on the World Heritage List prepared by UNESCO (Frey and Pamini 2009; Goodwin 2017) attract a large number of additional visitors (Landorf 2009; Su and Lin 2014; UNESCO 2019). Although there are obvious positive consequences to being included in the UNESCO-List, there are also negative effects (Frey and Steiner 2011; Martinez 2019; Vecco and Caus 2019). A cultural site on the list becomes better known. This makes it a promising target for terrorists and other nations to attack and destroy in a military conflict. Cultural goods and symbolic places become objectives in warfare. Terrorists and soldiers willingly destroy archaeological sites, monuments, and sacral buildings and loot museums, archives, and libraries to harm the enemy.

A prominent example of this occurred during the First World War. German artillery repeatedly shelled the cathedral of Reims, built in the fourteenth century in the French region of Champagne. It is considered to be one of the most important gothic churches in existence. The cathedral had, and still has, a high symbolic value to the French because between the twelfth and nineteenth centuries their kings were enthroned there. The German artillery fire completely destroyed the wooden roof structure and the medieval glass windows and also damaged the front structure. The historical value of Notre Dame de Reims to France was certainly known to the German military commanders. But the attacks were made precisely because the cathedral had such great importance to the French. Fortunately, Notre Dame de Reims has been restored and has been included in the UNESCO World Heritage List. More than one million tourists visit the cathedral every year.

Precious cultural sites have also been destroyed in local wars. Thus, the Old Bridge in Mostar, Bosnia-Herzegovina, dating from the sixteenth century, was entirely destroyed in 1993 during the Balkan war. Again, the symbolic meaning was crucial. The bridge was considered a rare masterpiece of medieval architecture and had a high symbolic value because it linked Catholic Croatians, Muslims, and orthodox Serbs and thus represented a bridge between the East and the West. The destruction can probably be attributed to the Croatian Defense Board, whose representatives were sentenced to long imprisonment (also for other crimes) by the International Court of Justice in The Hague. The bridge was rebuilt in 1995–2004 and included in the UNESCO World Heritage List in 2005.

During the same war, the army of Serbia and Montenegro attacked Dubrovnik, a formerly independent city-republic called Ragusa, with artillery fire. More than 100 civilians died, and many buildings were damaged. The bombardment was partly inflicted because this Adriatic city is on the UNESCO World Heritage List and therefore is well known all over the world. Meanwhile, the damage has been repaired, and Dubrovnik is one of the most visited cultural cities in Europe and therefore subject to heavy overtourism. Figure 2.3 shows the beauty of this city, which has been overrun by mass tourism.

In 2001, Afghan warriors of the Taliban blew up the Buddha statues of Bamiyan, which were created from red sandstone between 510 and 550 AD. They were also included in the UNESCO List. Terrorist fighters of the so-called Islamic State destroyed most of the magnificent archaeological sites of Timbuktu, Mossul, Nineveh, and Palmyra (Clemente-Ruiz and Aloudat 2019), precisely for the reason that they were on the UNESCO

Fig. 2.3 Dubrovnik in Croatia. Source: Picture by Ivan Ivankovic on Pixabay

List, and they knew that their destruction would attract enormous media attention.

7. *Popularity through Film and Media*

   Cultural sites become known worldwide through famous films, TV series, and videos, and are therefore chosen as tourist destinations. The Croatian city of Dubrovnik became a particularly popular place because some scenes of the fantasy series *Game of Thrones* were shot there (Connolly 2019). So many visitors now overrun Dubrovnik that the city council wants to limit their number to 40,000 per day. The small Austrian town of Hallstatt has also become famous through the Disney film *Frozen* and as a result is now overwhelmed by a large number of tourists.

The reasons listed all facilitate the rapid development of cultural mass tourism but depend on various rarely mentioned assumptions. A further increase in world income is assumed. However, international trade wars or pandemics can slow or even reverse this development, as the global experience during the COVID-19 crisis suggests. Such a pandemic could potentially influence tourists' behavior in the long run. An inherent fear of large crowds could change tourism flows towards higher preferences for less-visited destinations, which would counterbalance the popularity of superstar destinations. Another potential consequence is a change in the means of transportation preferred, reducing the attractiveness of travelling by airplanes and cruises. This shift would restrict tourism to a more local level, endangering destinations within regions with a higher population density. Airfares may well rise again in the future because low-cost airlines are forced out of the market and because additional taxes have to be paid as CO_2 compensation to save the climate. Possibly so-called "flight shame" may also reduce international tourism, but this prediction is particularly controversial. The success of cultural mass tourism also favors offering alternatives. This is already the case in the People's Republic of China. For example, there is an Eiffel Tower, an Acropolis, a Heidelberg Castle, and other cultural sites in various sizes. The Window of the World theme park in Shenzhen exhibits more than 130 reproductions of the world's most visited tourist attractions. This local offer, including European cultural sites, might prevent some Chinese people from making a trip to Europe to see the original.

The expected sharp increase in cultural mass tourism will not occur if a major incident such as a war happens. Unfortunately, such crises cannot be ruled out. A major nuclear incident may well happen when one thinks of the conflicts between the two nuclear powers India and Pakistan, the Arab states with Israel and the United States, or the conflicts of China and North Korea with western-oriented states in the Pacific. If such wars broke out, the situation would change fundamentally because international travel would become more difficult and cultural sites would be damaged or destroyed.

Transmissible diseases and pandemics can also affect cultural overtourism. In spring 2020, COVID-19 led to strict restrictions in national and international tourism. For example, the Chinese government completely banned group travel to Europe and other continents. Travel within China has also been severely curtailed. Other countries have introduced similar restrictions. Many of them almost closed their borders to personal traffic completely. This is also reflected in air traffic. Between January 2 and March 30, 2020, air traffic at Zurich Airport fell by more than 90% (Flughafen Zürich 2020). It is unknown whether those who are denied a journey to cultural destinations as a result of a pandemic will only postpone it and travel later, or whether they drop these plans altogether. However, ever larger proportions of the population in China and other countries are becoming part of the middle class and can afford cultural travel. In this case, very likely international tourism will recover after some time, and the problems posed by overtourism will arise again. Even if cultural tourism does not increase in the future and only reaches the levels existing before the pandemic, this will not constitute a sustainable scenario for many destinations—neither for local populations nor cultural goods.

It is well known that forecasts are difficult to make. No one can predict for certain what will happen in the coming years and decades. The decline of economic activity in spring 2020, and the blocking of international travel gives us the opportunity to reflect on the advantages and disadvantages of excessive visits to cultural sites. In any case, we should consider a possible, and also probable, great increase in cultural mass tourism.

2.4   What Are the Effects of Cultural Mass Tourism?

In many countries, tourism is an essential sector of the overall economy: just think of Italy, Greece, France, Spain, Austria, and Switzerland. The same holds for less-developed countries, such as Croatia and Slovenia, and most countries in South America and the Caribbean Islands. Many people benefit directly or indirectly from the jobs created and the income generated (e.g., Belisle and Hoy 1980; Tosun 2002). As a result, investments in infrastructure are stimulated, especially in transport and the construction of hotels.

Restaurants and souvenir shops benefit directly from tourists. In recent years, this has also been the case for locals who live in tourist centers. They can rent out their apartment for a short period (often via Airbnb) and thus generate far higher incomes for themselves than what they have to spend on the rent. It therefore pays for many of them to move to the suburbs where rents are cheaper. Mass tourism also promotes the economy indirectly, for example, in the cultural and taxi industries. This has revived old industrial cities, such as Bilbao with its spectacular Guggenheim museum of the architect Frank O. Gehry.

These positive effects of tourism on economic activity, which for a long time have been rarely disputed, are in line with the traditional view of tourism. Today, mass tourism is of great importance for the economy and society. It is one of the most significant business sectors, on which a huge number of people depend directly and indirectly. If it is controlled and restricted, many parts of the economy and society will be negatively affected.

However, the positive effects of tourism have been questioned, at least partially, in recent years (Martín et al. 2018). Thus, the question remains unanswered whether the local population really benefits substantially from the money tourists spend. Undesirable side effects are the often rapidly rising house prices and thus rents as a result of an increased demand from tourists (Lundberg 1990; Weaver and Lawton 2001; Barron et al. 2018). The local population can hardly afford to stay in their original places of residence. The general cost of living rises, too, because many traditional shops close due to high rents and lower demand from the decreasing number of local residents. The same is true for pubs and bars, which mainly serve tourists and charge much higher prices than before. Many local residents are forced to move to the outskirts of the tourist resorts. This effect is very obvious, for example, in Venice. This wonderful and impressive city is shown in Fig. 2.4. Nevertheless, many inhabitants have left the historic city center; only just over 50,000 people still live there, whereas in the thirteenth to seventeenth century there were around 200,000.

In contrast to the rental of individual guest rooms, the rental of entire apartments on Airbnb has increased noticeably. This indicates a more commercial use of this service. Additionally, the number of individual landlords offering several Airbnb apartments has grown significantly. The income from rentals is therefore distributed more unequally. The Gini coefficient of Airbnb income in Venice also supports this view. The coefficient is 0 if the income is equally distributed, and 100 if it is extremely unequal. Compared to the Italian average of 36, the coefficient in Venice is 60, which indicates a considerably less equal distribution. This underlines the assumption that a few

Fig. 2.4   Venice. Source: Picture by Artheos on Pixabay

people benefit disproportionately from platforms such as Airbnb (Picascia et al. 2017).

The same negative effects apply equally to the center of Florence, where almost 20% of all the apartments are listed on Airbnb.

As a result, the actual residents will lose the sense of having a home (Sans and Quaglieri 2016). There is hardly anything other than souvenir shops in the neighborhood, and mostly tourists frequent the expensive restaurants and bars (Bellon 2018 for the case of Berlin, and Thani and Heenan 2017 for Florence and Venice).

At the same time, tourism brings benefits for the local population. In Venice, museums consolidated in the Foundation of Municipal Museums of Venice (MUVE) must provide free entrance to people born or living in Venice. The implication of no subsidies from the regional government for MUVE is that tourists pay for the locals' benefit. However, for the locals, it is difficult, and sometimes even impossible, to visit their own art sites under reasonable conditions. For example, the inhabitants of Rome have to line up like ordinary tourists in the long queues in front of St. Peter's Cathedral. The same applies to the entrance to the Vatican Museums (OECD/ICOM 2018).

Figure 2.5 shows the long queue of tourists wanting to visit the Vatican Museums. The waiting line is normally even longer than can be seen in this picture. If an online reservation for a museum is needed, specialized

Fig. 2.5 Queue in front of the Vatican Museums in Rome. Source: Picture by Bruno S. Frey

professional companies often have an advantage in securing tickets for customers whereas individual tourists and local residents have more problems.

The distribution of income and wealth can also be influenced negatively by mass tourism. The individuals and companies that benefit from overtourism are often located outside the cultural sites, sometimes even in other countries. The same is true for the additional jobs created, which are often given to people who come from outside and are employed at a lower wage than local residents would demand.

In addition to these unfavorable effects on local residents, the main cause for complaints are the *negative external effects* caused by cultural overtourism. These effects are not reflected in costs or prices (e.g., Richardson 2017; Singh 2018) and therefore do not directly influence people's actions and behavior. The most important of these are the loss of authenticity which is so much sought by cultural tourists, the increased stress and loss of identity of the local population, the overuse and partial destruction of cultural sites, the ecological costs produced by the huge number of visitors, and increased crime and inappropriate behavior by tourists. These negative consequences of overtourism are discussed here.

With overtourism, the atmosphere and authenticity that most tourists look for and associate with a cultural place are lost. In a survey conducted in March 2019, 95% said that an "authentic atmosphere" is particularly important to them; a slightly smaller share (91%) takes cultural sites and museums to be particularly relevant (IUHB 2019).

On the streets and in shops and restaurants, tourists only encounter other tourists. According to D'Eramo (2018), "a traveler is only a tourist who denies being one." This effect is obvious for Venice, whose canals are full of tourists riding gondolas, as pictured in Fig. 2.6, while there are no locals. This effect also occurs in smaller places such as Hallstatt. The individual providers of services for tourists do not take such negative external effects into account and are mostly not interested in them.

These harmful effects are a public nuisance because they can only be avoided by a collective decision. If individual providers such as shops and restaurants decide to respond less to the needs of tourists and instead cater for local residents, they will lose turnover and profits compared with their competitors. This problem can only be overcome if all providers of services for tourists agree to limit their offers in such a way that the local atmosphere is preserved. As experience shows, such an agreement is difficult to achieve. It contradicts the motive of the individual providers to produce as much profit as possible. Ways and means are constantly sought and found to undermine the agreements to individuals' advantage. Moreover, it is not clear exactly how the desired atmosphere can be maintained.

The local population loses its identity because, as a result of the huge tourist masses, it is no longer able to form a community and to identify with other residents. Tourists occupy the places where the locals want to meet, for example, popular bars and restaurants and benches in central squares. Conversations and the exchange of ideas among the local population are severely impaired. There can even arise hostility between locals and tourists. The feeling of home is lost. Because of this, many residents prefer to rent out their apartments through Airbnb and move to the outskirts of the city or village. In this way, cultural traditions are lost. An exception only happens if the local traditions can be commercialized because they are attractive to tourists. Some local traditions, such as festivities in honor of a saint, or marking the change of seasons, might be preserved because they provide income to locals—but they at least partly lose their character as they are predominantly catered for tourists while the local population participates less and less.

Fig. 2.6 Overused canals in Venice. Source: Picture by Kirk Fisher on Pixabay

The huge numbers of tourists lead to an overuse of cultural sites. In the case of Angkor Wat and Machu Picchu, the historic paths were so heavily worn that wooden planks had to be inserted (Larson and Poudyal 2012; OECD/ICOM 2019). This affects the works of art in a very negative way. For this reason, UNESCO has included Machu Picchu in the list of the most rapidly destroyed sites on the World Heritage List (Hawkins et al. 2009). Until recently, around 6000 people visited the ruins every day. Since June 2019, tourists may visit this cultural site for only 4 h and then have to leave (Kamm-Sager 2019). Another well-documented example of overuse is Westminster Abbey in London. The masses of visitors have abraded the marble floor so that it is heavily worn (Fawcett 1998).

The ecological burden in emissions, pollution, and waste are growing significantly. Noise is also increasing so that both local residents and other tourists feel disturbed. One would expect the air on the island city of Venice to be particularly good and fresh. However, as a result of emissions from cruise ships, which run their engines even when they are docked, the air quality in Venice is actually very poor (Abbasov 2019). The exhaust fumes also damage the historic buildings, especially the facades. As a result, Venice is the city with the highest lung cancer rate in Italy. The fine particles damage the residents' health by causing respiratory and cardiovascular diseases. The mortality rate has also risen. Moreover, cruise ships generate electrosmog with their radar systems. As studies show, these are also harmful to health as they may cause cancer (Reski 2013).

The large numbers of tourists result in an increase in the crime rate, and more people behaving inappropriately. A frequently observable example is that tourists sit on the ground in the immediate vicinity of important cultural monuments, eat their food, and leave waste behind. There is also an increase in organized begging.

This list illustrates that the negative external effects of overtourism are mani-fold and significant. State intervention must, therefore, be considered, but nongovernmental solutions are also possible. In this book, I propose overcoming overtourism by cooperation between private entrepreneurs and state authorities.

Before moving to my proposal to overcome cultural overtourism by creating Revived Originals, I take a brief look at the reactions of the residents and others affected by overtourism.

3

Reactions to Cultural Overtourism

3.1 People Are Protesting

The impairment of art sites by cultural overtourism has led to anti-tourist movements and social unrest (e.g., Seraphin et al. 2018b; Clancy 2019). This has happened mostly in Spain, France, and Italy and also in the Netherlands. These protests have been discussed extensively in the traditional and social media (Zeng and Gerritsen 2014) under the keywords "touristification" and "touristophobia" (Milano 2017, 2018; Milano et al. 2018; Peeters et al. 2018: 29–30; Hughes 2018; Zerva et al. 2019). They are particularly visible in cities such as Barcelona, Lisbon, and Venice.

– In Barcelona, resistance against the flooding of the city by tourists, especially on the Rambla, is particularly strong. In order to enforce the political impact, various organizations have been founded, such as the Assembly of Neighborhoods for Sustainable Tourism and the Network of Southern European Cities against Touristification (Milano et al. 2019a, b).
– In Lisbon, a social movement called Morar em Lisboa (living in Lisbon) has published an open letter deploring the excessive dependence of the urban economy on tourism and speculation in housing.
– In Venice, some residents of the city regularly protest against the passage of the huge cruise ships through the Giudecca Canal (Vianello 2016). Figure 3.1 gives an impression of the devastating impact of cruising ships on the Giudecca Canal. The historic buildings, including churches, appear small compared to the huge cruisers. The big ships (grande navi) radically change the whole atmosphere of Venice.

B. S. Frey, *Overcoming Overtourism*, https://doi.org/10.1007/978-3-030-63814-6_3

Fig. 3.1 Cruise ship in the Giudecca Canal, Venice. Source: Picture by Bruno S. Frey

This resistance is supported by the organization No Grande Navi (No Great Ships). It is a manifestation of the unrest of Venetians worrying about the harmful effects of those cruise ships. It held a nonbinding referendum in which almost 18,000 inhabitants took part. In all, 99% voted against allowing the passage of cruise ships through the Giudecca Canal. These ships tower above the ancient buildings and look like monsters. The beauty of the city is affected badly as a result. Moreover, because of the intense water pressure, they damage the city, which is built on piles, and pollute the air. There is also the danger of accidents that can cause considerable damage. Residents held a procession as a protest from the Rialto Bridge to the Town Hall, called Venexodus (Armellini 2016), which received much media attention. At the Biennale 2019, the British artist Banksy, whose identity is unknown, displayed several paintings to protest against the negative impact of cruise ships. This activity was widely commented in the classic and social media.

Protests only work under certain conditions and even then, only after a considerable time. To be successful, protests must either be able to mobilize a large part of the population or disrupt economic or social processes. In such circumstances, politicians and members of the state administration feel compelled to take action. However, the critical factor is the pressure that the profiteers of overtourism can exert politically. Various groups such as hotel and restaurant owners, the transport industry, and shop owners are often well organized and therefore usually have more influence than the less well organized or even unorganized groups or individuals that are negatively affected by overtourism. The latter usually operate in their community and therefore have little influence in centrally organized countries. This is the case, for example, in Venice. Because of the protests of the local population, the mayor's office would have banned the giant cruise ships long ago from passing through the Giudecca Canal to dock in the city. However, the competence for this decision lies with the central government in Rome, which is open to the interests of the well-organized international cruise lobby. One reason for its importance in the political process is the regional distribution of income. Only 25–50% of the income remains in the area. The rest of the revenue goes to international cruise lines (Brida and Zapata 2010).

Still, in Barcelona, which is also profoundly affected by overtourism, demonstrators have achieved small successes. In 2015, Ada Colau, a mayor critical of tourism, was elected. Since then, she has been taking action against Airbnb's advertising for unlicensed apartments, among other measures. Other demands of the demonstrators, such as a ban on new hotel buildings in the center and a moratorium on new tourist apartments, have been passed by parliament in recent years (Hughes 2018). This shows that even though a public movement faces many obstacles to having its voice heard, let alone its demands implemented, achievements in favor of local residents are possible. But even more importantly, these various anti-tourism movements illustrate the severity of the overtourism issues affecting the residents.

3.2   Shifting to So Far Little Visited Cultural Sites

Nevertheless, there still are cultural sites that are not crowded yet. Reorientation to these is possible if sufficient information about "forgotten" places is available and if they can be visited with little effort. But this is rarely the case.

A question consequently arises: what are the drivers that make a place popular or prominent? If popularity is fostered by digital means, such as social media, new superstar destinations are created every time a site is promoted. This danger arises because of the broad reach of particular influencers or information sources. For instance, National Geographic Travel has almost 40 million followers on Instagram, and the main site of the magazine *National Geographic* even has 143 million followers, who are regularly updated with new recommendations for destination choices. These are only examples, but if social media continues to strengthen its influence on destination choices, the effect can run out of control.

So far, some cultural sites are not prominent tourist destinations because they are more difficult to reach and provide fewer amenities. There is often little tourist infrastructure such as inns and hotels. Sometimes the cultural sites are not accessible because they are closed to prevent vandalism. It therefore requires a higher effort for prospective visitors to go there. The more complicated search required collecting information and the higher transport costs tend to discourage visitors.

Improving the infrastructure of previously neglected art sites is certainly one way of dispersing tourist masses better. But one should not expect too much from this measure. In particular, it will only temporarily lessen cultural overtourism because some of these places will soon become overcrowded too, due to the huge number of tourists expected in the future. Since social and traditional media regularly recommend remote places as particularly delightful destinations, a concentration of visitors can also be expected as a result. This, again, causes strong negative overcrowding symptoms.

An example of this is the mountain restaurant Aescher on the Alpstein in the Swiss Canton of Appenzell. It was featured on a cover photo of the magazine *National Geographic* under the title "Places of a Lifetime." As a result, such a massive invasion of tourists took place that the long-term tenant couple decided to quit because they could no longer deal with the huge number of tourists (Kamm-Sager 2019). A massive and unexpected influx of tourists to a destination has fatal consequences because the infrastructure and other needed facilities are missing.

3.3 Government Intervenes

Central, regional, and local governments are making efforts to control and manage overtourism. The four most frequently adopted measures are bureaucratic regulations, attempts to distribute the masses of tourists better across time and place, expanding the capacity of destinations, and increasing the cost of the visit (Peeters et al. 2018: 102). Other forms of interventions are also being considered and sometimes introduced (McKinsey & Co and World Travel & Tourism Council 2017; OECD/ICOM 2019).

3.3.1 Information and Appeals

Tourists may be advised to visit the cultural sites at times when there are fewer visitors. They may be informed in real time about how crowded certain cultural sites are and especially about the waiting times for admission to be expected. This information and advice can refer to the season, the day of the week, or the time of day. Addresses for such information are not only important for individual tourists but also for all travel organizations.

The city of Venice has launched an awareness campaign called #EnjoyRespectVenezia to reduce the negative effects of overtourism. It recommends 12 rules of conduct. Including suggestions to explore lesser-known places or visit Venice at a less busy time (Città di Venezia 2019). This campaign was launched some years ago (2017) but had little success in countering the rising numbers of tourists. Within the same campaign, an online projection of the expected number of visitors to Venice has been provided and marked with the colors of traffic lights (Città di Venezia 2020).

Individual institutions might be able to provide proper information and appeals, but coordination with other institutions in the same city is more challenging and administratively costly.

It is questionable whether such advice has much of an effect. Most likely, visitors will not adapt fully to the change in visiting times. If art lovers have to queue for one or more hours in front of a famous museum to be admitted, both, individual and group tourists will try to avoid this inconvenience. However, a bizarre effect could occur. If too many potential tourists choose a visiting time recommended as being less crowded, this new time slot will also become crowded, and the waiting time will be reduced only slightly or not at all. The reduced congestion at the former peak times lowers the implicit visiting costs and will attract additional tourists.

In general, appeals have little impact on human behavior. They are not binding, and those who do not follow them face few or no disadvantages. In contrast, government measures that impose a cost on undesirable behavior are more effective.

3.3.2 Marketing Efforts

Various alternative measures can be taken to try to influence the distribution of tourists over the year and thus extend the season, although weather conditions do play a role. Another option could be to distribute tourists more widely over an urban area. In Berlin, for example, there are efforts to divert visitors from crowded sites such as the Brandenburg Gate and the Pergamon Museum to Köpenick, Reinickendorf, and Spandau. And Amsterdam promotes a castle outside the city with the name Amsterdam Castle. Even nearby beaches are named Amsterdam Beach. These marketing efforts are undertaken to reduce the pressure of tourists in the city centers. However, it is highly doubtful whether these measures will succeed.

Access to a cultural site can also be better managed by online registration. This procedure is already used by many museums. However, it discriminates against individual visitors because professional tour operators have a better chance of obtaining admission for their customers. Furthermore, the total number of visitors is only slightly affected by this measure. It also has the effect that those who have not bought tickets in advance are annoyed with the longer queues, which gives the place a bad reputation.

Advertising for visiting crowded cultural sites can be restricted or omitted. Amsterdam no longer advertises the city as a tourist destination. By advertising other places worth visiting, cultural tourists could be encouraged to explore such new areas. Still, such provision of information is only effective if the alternative sites can be reached with little additional effort and if they are equally attractive. How influential the city's advertising campaigns will be in determining tourists' choices of destination is questionable, compared to other information sources, such as user-generated content (UGC) on social media. If instead, the number of culturally oriented tourists continues to increase as strongly as in the past, these new destinations will soon be as crowded as the well-known ones.

3.3.3   Temporal and Local Administrative Restrictions

In many places affected by cultural overtourism, visits are regulated in several ways (Martín-Martín et al. 2019). Some sites can only be visited at a strictly regulated and limited time. As already mentioned, access to the Peruvian ruins of Machu Picchu is restricted to 4 hours.

Some places are even completely closed to tourists, such as the caves of Lascaux and Altamira with their impressive prehistoric paintings. Instead, new caves have been created and the paintings precisely replicated.

In addition to such formal bureaucratic and pragmatic interventions, cultural tourists are guided on the spot, almost like a herd of cattle. In the Vatican Museums, on some days, 30,000 tourists want to see the Capella Sistina with the famous paintings by Michelangelo. Consequently the visitors are pushed through the museums without having the opportunity to admire the wonderful paintings of Raffael, among others, in the *Stanze*. There is another queue in front of the *Sistina*. Once in the chapel, the guards hush the visitors through with harsh words and gestures. The same applies, for example, to the situation in the Louvre, which I have also experienced personally. A large part of the tourists want to see the Mona Lisa. Figure 3.2 shows that most visitors in front of Da Vinci's masterpiece want to take a picture—as if this presumably

Fig. 3.2 Tourists in front of the Mona Lisa in the Louvre. Source: Picture by Thomas Staub on Pixabay

best-known painting in the world had not been photographed before and as if it were not available on various online platforms such as the Web Gallery of Art. The tourists can only stay in front of the picture for a very short time; they are driven onwards by the staff.

The crown jewels in the Tower of London can only be admired for a very short period because visitors have to mount a treadmill that allows each person the same amount of time to look at the collection. Such interventions in the stream of culturally interested tourists are questionable. Not every person shows the same interest in a cultural object and wants to proceed equally quickly. Some wish to examine an object for a long time and from many different angles.

Another way of dealing with the masses of visitors is to restrict or even ban short-term rentals of apartments. One option is to impose bureaucratic hurdles on this kind of rental. Local residents then have less incentive to leave the cultural centers. This strategy interferes with private property rights, which is questionable. However, it also makes a city more attractive for visitors because they meet not only other tourists but also local residents.

Administrative intervention is also possible on the supply side. The infrastructure can be improved to cope better with the masses of visitors. Thus, additional parking spaces can be provided in suitable locations, thus limiting search traffic, which causes noise and exhaust fumes. However, the improved infrastructural conditions are likely to attract even more tourists.

3.3.4   Tax Incentives

Levying taxes of various kinds can influence tourist flows. For example, tourists who stay on the outskirts of Amsterdam pay a lower tax than those who want to stay in the center. But these differences in the tax burden are only effective if they outweigh the disadvantages caused by evading them. Those who choose a hotel outside Amsterdam have higher transport costs and need more time to travel to the attractions in the city center.

3.3.5   Price Increase for Visitors

Overtourism can also be influenced by directly or indirectly increasing the price of a visit to a cultural site so that it becomes less attractive to tourists. Such an intervention can be justified because of the negative side effects caused by overtourism. For example, parking fees may be raised for private cars, but above all for tourist buses. Of course, the risks of unintended consequences have to be taken into account, as the higher costs should affect the tourists but not the local population. This can be achieved in various ways, such as by annual parking fees for the local population or a special offer of parking facilities for tourists. Cities visited by cruise ships can increase the berthing fees for ships in the port.

Moreover, the costs of visits can be raised, and thus the influx of visitors reduced, by increasing the government fees for overnight guests. Additional taxation of hotels and restaurants is also an option that can be extended to all providers of other goods and services.

Higher taxation of the tourism sector will cause resistance from suppliers because their profits will be reduced. It cannot be assumed that the additional taxes can be passed on to consumers in full. Nonetheless, the distorting effects of additional taxation are equally important. Day tourists are not affected by a higher overnight tax. Passengers on cruise ships receive meals onboard and therefore spend little money on restaurant visits. In most places, even overnight stays at Airbnb are not subject to taxation. This may change in some places in the future, but there are many ways to avoid such a tax.

Visits to cultural sites can be made more expensive directly by charging an entrance fee. This measure is efficient because those who are responsible for the overtourism have to take responsibility by paying more to view and experience cultural assets.

An entrance fee can take many forms. In most cases, an adult person pays a certain price and children pay less. The price of admission can be varied according to occupation and age (students, soldiers, and pensioners pay less) or citizenship (entry tickets for residents of the country or members of the European Union are cheaper), as is the case with admissions to many other cultural events such as theater and opera.

The most suitable solution is a variable admission price based on the existing excess demand, named peak load pricing. If a particularly large number of people want to visit an art venue at a specific time, the price is raised; if a smaller number of people want to do so, the price is lowered. When a cultural site is not filled with visitors, no entrance fee should be charged. In practice, however, it is not easy to determine the price for visiting a cultural site. Potential visitors may consider a price that changes constantly according to excess demand to be arbitrary. In fact, some surveys have shown that prices based on excess demand are not well accepted by the public because they are considered unfair (Kahneman et al. 1986; Frey and Pommerehne 1993). However, more recently, people have experienced peak load pricing for instance when buying airplane and train tickets and have gradually become accustomed to prices that vary over time.

The hotel industry has long since applied dynamic prices. As more business moves online, peak load pricing becomes easier to implement. The same is true for cultural events. They often offer seats at a lower price on a normal weekday than on Fridays or Saturdays. Another example is that of remaining tickets, which are allocated at a lower price shortly before an event (Metz and Seesslen 2019).

Social aspects can also be considered in other ways. One option would be to treat tourists differently depending on their nationality. A person who comes from a country with a high average income has to pay more than a person who comes from a developing country. Again, there are many ways to evade such pricing, especially on the part of professional tour operators. This can lead to injustice. The substantial administrative effort and cost needed must also be taken into account.

Cultural tourists coming from far away could be given preferential treatment because they show that the visit is particularly important to them. However, the fact that the entrance fee of a few euros represents a minimal proportion of the total cost of a trip and is therefore negligible speaks against this. For example, those who travel from China will not be deterred from visiting Venice because of a fee of 3 or even 10 euros (FAZ 2019).

Charging entrance fees only makes sense if access can be controlled with little effort. This is obviously the case with museums. Most museums in the world charge an entrance fee. However, 50 national museums in the United Kingdom have been granting free admission to permanent exhibitions since 2001. The London museums are mostly visited by tourists and are among the most cherished and popular attractions in the United Kingdom. Admission to the permanent collections of the National Gallery, the Tate Gallery, the New Tate, the British Museum, and the Victoria and Albert Museum is free of charge. Only for special exhibitions is an entrance fee charged.

An admission fee can also be charged at individual cultural sites. For example, a visit to the Taj Mahal costs just over 12 euros, while Machu Picchu costs around 40 euros.

The Austrian village of Hallstatt is flooded by tourists and receives almost 20,000 coachloads of tourists per year. Figure 3.3 shows that it is indeed placed in a lovely setting on a lake and with impressive mountains nearby. In order to cope with the large number of tourists, the bus companies are given a time slot between eight in the morning and five in the evening. The buses must stay at least two and a half hours in the allocated parking lot and have to pay a fee of 80 euros. This is an attempt to reduce the number of buses to 8,000 per year (Benz 2019). The municipality can thus generate considerable income, and the turnover of the restaurants will increase.

Fig. 3.3   Hallstatt, Austria. Source: Picture by Arvid Olson on Pixabay

The Catholic Church wants to avoid the perception of its churches only as museums and works of art. In order to generate some income, the most frequently visited churches charge an entrance fee for easily identifiable areas, such as chapels, the sacristy, and the cloister. In Venice, tourists may visit several famous churches with an overall ticket, but local worshippers who visit the churches for prayer are exempt.

For culturally important cities, such as Rome, Florence, Salzburg, Bruges, and Prague, no entrance fee is charged because these cities have many access points that are difficult or impossible to control. Venice is a major exception as nearly all tourists except cruise ship passengers arrive at the Piazzale Roma by train or car and travel to the city center from there. The City Council of Venice plans to introduce an admission fee of between 3 and 10 euros to limit the number of tourists (Giuffrida 2019). An entrance fee should also be possible for Dubrovnik, because the old town can only be entered through two main gates, which are easily controllable. However, even under favorable conditions, considerable administrative effort is necessary as it must be determined who is entitled to free entry. Does this apply to regular or even one-off suppliers to the restaurants and shops? And what about people who rent an apartment for some time? Should the admission price also apply to Airbnb guests?

Entrance fees are also charged for scenic attractions. For example, the island of Komodo in Indonesia has so far charged around 9 euros for admission. Now a price rise of up to 450 euros is under discussion. The reason is the rapid decline of a species that lives only there, the Komodo dragons (Marti 2019).

3.3.6 Auctioning of Visiting Rights

Another approach is to limit the number of admissions to a cultural site so that visitors and the local population hinder each other as little as possible. Those who absolutely want to visit a certain place and are willing to spend enough for this purpose will pay a high price for the visiting right. In contrast, people who are only moderately interested will not want to spend much money on a visit. They will therefore only be able to visit the place if few other tourists want to do so. Auctioning the visiting rights thus helps ensure that those who benefit most from a visit are able to enter.

An auction procedure has some disadvantages. It is quite difficult to implement the administration necessary in place. Applicants with higher incomes have an advantage. Auctioning entry tickets can be rigged under some conditions. The participants at the auction may be able to form a coalition, popularly termed an auction ring, by which bidding is reduced and therefore the entry price is low. The successful bidders can then sell the entry tickets at a higher price and create an undesirable secondary market. The profit made is then distributed among the members of the coalition. The auctioning authority should be aware of such a risk. It must therefore make an effort to prevent such profit-seeking coalitions. This is not easy to achieve as such coalitions are formed secretly and are therefore difficult to detect.

An auction of access rights therefore would only be considered fair in very few cases.

3.4   Visitors Are Banned

The measures discussed so far to curb cultural overtourism all share at least one major flaw: Some people who want to admire the sites are excluded. This is particularly evident in the case of administrative restrictions. Those who do not fit in with the temporal and local conditions will not be allowed to visit the cultural treasures they are interested in. If an entrance fee is charged, all those who do not wish or are unable to pay the required price are denied the visit. This restriction is likely to affect families with children in particular. For example, although the entrance fee planned for Venice is low at 3 euros for a single person, a family of five must pay 15 euros. As the price is likely to be raised in the future—a fee of as much as 10 euros is being considered—the sum for a whole family can be considerable.

Bearing in mind that governments subsidize most cultural institutions in part to enable less wealthy people to enjoy these cultural goods, it seems paradoxical to increase the entry prices. The links between government subsidies, the role of museums, and more stakeholders need to be considered.

If culturally interested tourists visit other art sites that have been visited rarely so far, the negative effects are less damaging. The tourists are compensated for their "loss" by at least being able to visit these other places. Whether they value these sites just as much as those that have been denied to them remains open. It is also possible that these sites will also be overrun as a result.

Restricting access to cultural sites by government regulations or pricing has another major drawback. The excess demand created by these measures—more people wanting to visit a cultural site than the measures allow—offers opportunities for corruption. Interest groups and individuals can offer money to the administrators controlling entry to allow a visit in excess of the number allowed. Excess demand can also be exploited by establishing a black market. If entry is restricted by government regulations, for instance by offering entry tickets on the internet, enterprising people will use all sorts of tricks to obtain as many of them as possible. Because they are valuable to potential visitors, they can be sold at a price, thus making a profit.

The same procedure can be used if the entry fee is too low to bring demand and supply for entry to the cultural site into equilibrium. This is likely to be the case: suppliers are reluctant to impose high prices because they fear being accused of behaving in an antisocial manner. In such a case, there is again an excess demand, which can be exploited on a black market. Agents buy tickets at the regular price and then sell them at a higher price to organizations and individuals wanting to visit the site.

Part II

Overtourism: A Radical Proposal

4

A Positive Alternative: *Revived Originals*

I propose a radically different approach to address the pressing problem of cultural overtourism. Instead of limiting demand, which would exclude many art lovers, *supply should be increased by providing alternative offers*. I call these additional new cultural sites Revived Originals. By creating a broader range of sites, many more tourists are able to enjoy culture. At first sight, the proposal to expand the supply of cultural sites seems absurd and has been declared in the literature to be impossible (e.g., Smeral 2019). Cultural sites have long been considered to be historically determined and cannot be increased. But after taking a closer look, the idea of expanding the supply proves to be both reasonable and feasible.

Revived Originals have the potential to be experienced by a large number of tourists. They are entertaining and offer new insights to culture both on the intellectual and the emotional level. People who absolutely want to see the original site can enjoy it under less stressful conditions because the number of visitors will be lower there, thanks to the alternative supply of Revived Originals. The massive stream of tourists to the original sites will be reduced. For many tourists, the new supply of Revived Originals will be welcome. It will allow them to immerse themselves in the art and history of a cultural site and enjoy additional benefits by doing so.

B. S. Frey, *Overcoming Overtourism*, https://doi.org/10.1007/978-3-030-63814-6_4

4.1   What Are the Features of Revived Originals?

4.1.1   The Most Important Buildings Are Replicated Identically

Experience has shown that from a tourist's point of view, a few buildings define a cultural site. A large share of the tourists who visit a city concentrate only on the three or four most important monuments they know from the media.

Venice is a good example. Many tourists focus solely on the Doge's Palace, St. Mark's Church, St. Mark's Campanile, the Rialto Bridge, and those parts of the Grand Canal that are visible from there. These sites are often captured in selfies, with tourists' backs turned to these buildings and sites. Afterward, these tourists move quickly to another cultural site, which is documented in the same way for friends and followers. A 15-second video is often regarded as the major goal of the journey (*The Economist* 2019). A closer look and an approach to a work of art are either not intended or are made difficult, or even impossible, by the masses of visitors.

Therefore, it can be useful to copy the most frequently visited art sites and place them somewhere else as Revived Originals. Identical replication is easily feasible with today's technical capabilities; visitors cannot tell the difference from the original.

4.1.2   Digital Technology Is Extensively Applied

The buildings and works of art to be reproduced must be enhanced by modern digital technology. All technical capabilities and in particular virtual technology can be extensively used to achieve this goal (e.g., Aichner et al. 2019; Daponte et al. 2014).

Computer-generated objects (i.e., augmented reality) can visualize real buildings and other sites. The scenes created in this way look so real that they cannot be distinguished from those in the real world. The original and digital objects coexist simultaneously and are merged. These Revived Originals are three-dimensional (3D) and can be experienced in real time. In recent years, technological innovation has led to great progress in augmented reality. It can be expected that this development will continue in the future. The difference between the original and the augmented reality is therefore becoming ever smaller and will gradually disappear.

Augmented reality provides visitors with the opportunity to dive into a three-dimensional environment created by computers. They can interact with the Revived Originals, creating a virtual reality. Visitors can select the aspects that suit them most and interact with them. All five senses are used, including, for example, the smells, sounds, and lights typical of a city or region. Soon, it will no longer be necessary to put on an inconvenient virtual reality helmet. Instead, in the future the feeling created can be generated by invisible means.

Digital technology enables visitors to immerse themselves in cultural sites and thus experience them directly and intensely. Digital artistic artifacts can take various forms (Arnold 2008). For example, they can use videos, 3D copies, digital overviews through digital surveying, in which distances are recorded electronically and placed in an overall context, radar systems, which can be applied to look below solid surfaces, the global positioning system (GPS), heat sensors, and acoustic measuring systems.

Using the most advanced digital technology can be viewed as an extended form of "scenography" (see, e.g., Howard 2009; McKinney and Butterworth 2009; Aronson 2018). This practice enhances an original creation by using a coordinated synthesis of space, sound, light, environment, and costume design. It not only provides an intellectual but also a sensory and emotional experience. Scenography has so far mainly been used in theater design and museums to give the viewers a better understanding of the exhibits. Revived Originals represent a large extension of this approach as it is applied to all kinds of cultural heritage, including complete historic cities and even a combination of such cities.

The Venice Time Machine is a digital project designed to allow people to dive into the past of Venice. It is considered a leading project in the digital world. The gigantic State Archive of Venice contains many sources from over the centuries that are being used to reconstruct the complete history of the lagoon city. This approach allows visitors to stroll through the city while looking at it in different epochs and, for example, watch craftsmen building the Doge's Palace. The course of epidemics and the workings of the Venetian financial market can also be experienced in this way. However, time will tell whether the Venice Time Machine project can be completed successfully. Disputes between the parties involved, the University EPFL Lausanne and the State Archives of Venice, about the methods and prospects of success of digitization jeopardize this project (Hafner 2019). Nevertheless, it demonstrates how digital methods can be used to bring a Revived Original to life.

A good example of how demolished art sites can be reconstructed by virtual reality techniques was shown by an exhibition at the Institut du Monde Arabe in Paris in 2018–2019. The Bundeskunsthalle in Bonn later adopted this exhibition. Large projections guided visitors into the heart of the three legendary ancient metropoles of Mosul, Aleppo, and Palmyra, which were destroyed by Islamic terrorists, and Leptis Magna, which was severely damaged for other reasons (Clemente-Ruiz and Aloudat 2019). With virtual technology, the universal heritage of the world is not completely lost even after destruction. The technology to create digital artifacts is already quite advanced and will improve further in the future (Greengrass and Hughes 2008).

Holograms can also be very useful. When they are designed professionally, it is hardly possible, or even impossible, to tell whether a real person or just the hologram of a person is actually standing in front of you. This technology can bring historical events to life. An example would be the legendary activities of Giacomo Casanova as a womanizer in Venetian society in the eighteenth century: his sensational escape from the lead chambers of the Doge's Palace, his restless travels to many royal houses in Europe, and finally his last years at Dux Castle in the Kingdom of Bohemia can be depicted in an exciting and instructive manner. Likewise, the lives of famous composers, painters, writers, statesmen, and generals, and of course the lives of famous women can be shown, such as the official mistress of the French King Louis XIV, Madame Pompadour in Versailles. Even more fascinating is the Swedish Queen Kristina (1626–1689), who converted to Catholicism. Her story could portray life in the city of Stockholm at that time, which she turned into the "Athens of the North." Queen Kristina succeeded thanks to close contacts with some leading scientists of her time. Her life in Rome after stepping down from the throne could also be visualized for visitors.

Holographic augmented reality is, as the name suggests, a combination of augmented reality and holograms. This technology uses a headset to provide the user with a 3D experience. Visitors can even move artifacts in virtual space with gestures. This interactivity allows a more in-depth and more exciting examination of cultural and historical artifacts. TombSheer is such a Holographic augmented reality application that offers an insight into the Egyptian Tomb of Kitines at the Royal Ontario Museum. The formerly passive observer can now actively participate in the experience. The combination of acquiring information and movement increase the visitors' learning (Pedersen et al. 2017). This creates a valuable opportunity for educational institutions to link prominent events in an entertaining form with knowledge generation.

Digital twins go one step further than holograms. They represent more than a mere copy of an original. These twins take on a life of their own and are connected to their originals by a variety of sensors. They are constantly evolving, creating a flood of new data (*The Economist* 2020b). Similar to artificial intelligence, they can recognize objects and faces, languages, and even the smells encountered by visitors to cultural sites. Interactions with tourists can also be realized, enhancing the visitors' experience. This adds a large number of additional possibilities for Revived Originals in the future.

The multimedia exhibition Van Gogh Alive, which has been shown with great success on six continents and in 130 countries so far, demonstrates in an impressive way how modern digital technology can be used. The work of this unique painter, already known to most people, is presented to visitors by a combination of light, pictures, and music. Although the exhibition does not present any original paintings or even copies, the works of van Gogh are brought to life and woven into a story via 360° projections. They make the audience believe that they stand right in the middle of his paintings. In addition, they can learn about sources and experiences by which the painter was inspired. The visitors become part of the events.

The exhibition presented in the Louvre of Paris on the occasion of the 500th anniversary of Leonardo da Vinci's death in 2020 not only showed some famous paintings created by this genius but also tried to convey his entire oeuvre to visitors. Various masterpieces shown in other museums, such as the *Madonna with the Carnation, Cecilia Gallerani*, and *Ginevra de' Benci*, could not be exhibited because they could no longer be transported. However, they were on display as infrared images in their original size. Only the 39th painting in the exhibition was an original painting by da Vinci. Nevertheless, the exhibition was a huge success; there were over a million visitors. According to the museum, this is a historical record. Never before has an exhibition attracted so many people interested in art. This shows the vast possibilities of using digital technology to create replicas of artworks, which millions of visitors accepted.

4.1.3   Art Sites Are Historically and Culturally Embedded

Both the historical development and the artistic significance of a cultural site can be shown. Whoever visits a Revived Original receives a close and exciting impression of the cultural site. It differs significantly from the experience of the many travelers who visit cultural sites only to shoot selfies but have little or no knowledge of their history or artistic significance.

Visitors to Revived Originals can be presented with various options according to their interests and educational background. Children may be attracted by digital versions emphasizing aspects suitable to their age; other presentations may focus more on history or specific cultural aspects such as music, paintings, or literature. This may provoke interesting discussions between the different kinds of visitors, for instance, between children and their parents who have experienced different aspects of a cultural site. Such differences increase both the enjoyment and the benefits of a visit to Revived Originals, which gives them an advantage over the originals.

Venice can serve as an example in which the creation of a Revived Original may be most fruitful. It is reasonable to assume that only a few of the 130,000 people who visit Venice on some days know much about the history and cultural importance of this lagoon city. Therefore, the replica buildings and digital technology used in the Revived Original of Venice must mirror the original site closely. Digital techniques can manage the transition from the replica buildings to the environment corresponding to the original. For example, the view of the Grand Canal from the Rialto Bridge can be displayed virtually. Visitors will find it difficult, or even impossible, to determine where the replicas end and where virtual representations are added. Visitors should also feel as if they were really present at the time when the cultural sites were created and developed. The culture shown must be displayed in a way that is easy to comprehend. The historical and cultural experience conveyed provides visitors of a Revived Original with an additional benefit that is sadly missing when visiting the overcrowded original site. This experience distinguishes Revived Originals from the original places that are negatively affected by cultural overtourism.

4.1.4   Combining Copy and Digital Technology

The importance of exactly replicating buildings, and how the various dimensions of virtual technology can be used depends on the specific cultural site. Without augmented and virtual reality, Revived Originals are similar to the buildings of existing theme parks such as the various Disneylands and the Europapark in Rust, Germany. One could even go one step further and imagine Revived Originals where no buildings at all are replicated and the relationship of the site to history and art is accomplished solely through digital technology.

In recent years, museums have engaged in both real extensions of their space and digital means to present their holdings. An example of the former strategy is the London Tate Gallery, which was extended to the Tate Modern. It is situated in a different place and has been hugely successful in attracting attention and numbers of visitors. In more cases, existing museums have extended their space by adding new wings (such as the Swiss National Museum in Zurich) or adjacent buildings (such as the Kunsthaus in Zurich). Typically, famous architects such as Jean Nouvel, who designed the Louvre Abu Dhabi in the United Arab Emirates, undertake such extensions.

Museums have also used digital technology extensively to attract attention and possibly additional visitors (scenography). From the first, very simple websites in the 1990s, museum websites have developed enormously and became a central pillar of these institutions (De Lusenet 2007). Meanwhile, in addition to information about prices and opening hours, many museums offer online catalogs or recently even virtual tours through their exhibitions. Virtual techniques are used to enhance and complement the museum experience through interactivity and richness of content, thus making them more immediate to visitors. For example, the Louvre in Paris can be visited online. This opportunity has been used in particular for the presentation of special exhibitions. Due to the lockdown imposed by governments to curb the COVID-19 virus, this possibility has recently been used to a far greater extent than before.

It may be debated whether a virtual presentation of art objects raises people's interest in seeing the originals or whether the effect is substitutive: people no longer find it necessary to visit the actual museum. It is also an open question what benefits people derive when they see art objects in reality or virtually. It may well be that in the latter case people can better focus on those exhibits in which they are really interested.

It is not only museums that can benefit from using advanced digital technology but it is also possible to present specific landscapes in this way. In Neuhausen am Rheinfall near Schaffhausen, Switzerland, artist Beat Toniolo created a 360° multivision show of the Rhine Falls. This kind of presentation is new to Switzerland or elsewhere in Europe. Over several years and covering all four seasons, millions of images of the Rhine Falls have been collected. The images are complemented by music, light shows, and natural sounds. The aim is to bring the Rhine Falls to life in a way that has never been seen before.

These examples show that art can become part of a virtual world. Revived Originals that exist solely in virtual form are a special type. They offer innovative possibilities for presenting culture and history and convey them more effectively to people. One could even argue that a digitally created Revived Original is more "real" than the original art site because of the many additional features.

4.2   What Are the Advantages of Revived Originals?

In contrast to today's usual visits to overcrowded cultural sites, Revived Originals have several great advantages.

A much closer relationship with the visited cultural sites is established. Tourists do not just tick off the highlights recommended on the internet and then travel on. Instead, both the history and the cultural connections are conveyed to them in an impressive and attractive way. The visit becomes a broader and more rounded experience and the tourists feel more emotionally connected with the art sites.

The stay becomes much more pleasant. From the very beginning, the Revived Originals are planned so that visitors can not only immerse themselves in the culture but also do so in an enjoyable way. The large number of tourists can be distributed sensibly among the various attractions. This eliminates the haste and crowds that characterize today's sites of overtourism. At the same time, amenities such as restaurants, hotels, and souvenir shops are easily accessible and provide clean toilets. Proper facilities can also be offered to people with disabilities. This includes the infrastructure they need and implies, for instance, that the whole premises are wheelchair accessible. This issue is much neglected today in many historical sites, often for obvious reasons. It is difficult, or impossible, to make present-day Venice or Macchu Piccu fully accessible to people with disabilities.

The time needed for a visit is considerably reduced because long queues and long waiting times are eliminated. Thus, today in Venice, tourists must often queue for a considerable time before being able to board a vaporetto to get from the Piazzale Roma to the city center. When different cultural attractions are combined in one place—such as the northern Italian cities of Verona, Siena, Pisa, Padua, Bergamo, and Vicenza—the time spent on cumbersome transportation during a visit is reduced even more.

Revived Originals protect tourists from crime and also from commercial begging. This eliminates one of the worries that families with children have in particular.

Revived Originals are built according to the latest environmental and sustainable standards. Pollution is reduced to a minimum by the most advanced technology. The advances in zero-carbon buildings are noteworthy and support the environmental sustainability of Revived Originals. The main measures to achieve this are the use of renewable energy sources, the right construction materials, and also reducing energy demand by improving efficiency. Bearing in mind that buildings contribute about 45% to total carbon emissions and energy consumption worldwide (Butler 2008), Revived Originals have great potential to contribute to the reduction of carbon emissions associated with buildings and tourism.

It is often the case that many constraints restrict the widespread use and implementation of zero-carbon buildings. These boundaries include geographic, density, institutional, and even stakeholder limitations (Pan 2014). However, these limitations are often associated with dense urban settings or improving existing buildings and not new constructions. Therefore, Revived Originals could leapfrog at least some of the problems that often arise.

The easy access and the assembled sites obviate further environmental harm. The long search for parking spaces by those arriving by car or bus, which causes noise and fumes, is no longer necessary because sufficient parking space is provided.

Due to the reduced number of visitors at the original sites, local residents there suffer less from the pollution of air and water, criminal acts and annoying beggars, noise, congestion of sidewalks and streets, and long queues.

Revived Originals also contribute to the preservation of cultural sites to posterity. Important buildings and art objects are copied one to one. Although original works of art may decay because they are maintained and preserved insufficiently, damaged by environmental impacts, or destroyed by wars, terrorists, or natural disasters, they will still be there to experience for future generations. Since a part of cultural mass tourism will be redirected to the Revived Originals, the impact on the original cultural sites can be reduced, and they can be maintained in a better condition.

This enumeration shows the multiple advantages of Revived Originals over today's excessive cultural overtourism.

4.3   Similar Sites Already Exist

4.3.1   Prehistoric Paintings in Altamira, Lascaux, and Chauvet

Cantabria in Spain and the Dordogne in France each possess a cave with exceptionally beautiful prehistoric images of deer, bison, horses, wild boar, and also people: the caves of Altamira and Lascaux.

In Altamira, the paintings are generally thought to be about 15,000 years old, but there are also estimations that they were in fact created 36,000 years

Fig. 4.1 Prehistoric paintings at Altamira. Source: Picture by Welcome to all and thank you for your visit on Pixabay

ago. Figure 4.1 illustrates how marvelous the prehistoric wall paintings are. For a long time, tourists were allowed to visit these caves. However, the number of visitors rose to such a degree that the wall paintings were considerably damaged by the humidity resulting from the visitors' breathing and sweating. Therefore, in 1977, it was decided not to let any more tourists into the original caves. As compensation, a facsimile was displayed in the visitor center, which was built about 500 meters away from the historic cave. The original cave was integrated into the comprehensive museum area and the replica was opened in 2001.

Because the paintings in Lascaux were infected by mold as a result of the many visitors, an exact replica was built only 200 meters away (Lascaux II). A traveling exhibition was also designed (Lascaux III). The cave was replicated even more elaborately in Montignac, also situated in the Dordogne and opened to the public in 2016 (Lascaux IV).

The reconstructions of Altamira and Lascaux have been so successful that even researchers study the copied murals because they can be examined in greater detail. Tourists have readily accepted these new caves with the copied paintings. Indeed, from the opening of the Altamira Museum in 2001 up to 2019, on average more than 260,000 people have visited each year.

The Chauvet-Pont-d'Arc Cave in Southern France was discovered in 1994 but closed to tourists soon after because its prehistoric paintings suffered from the many visitors. The reconstruction called "Cave Chauvet 2—Ardèche" was opened in 2015. Various technological methods were used to create a satisfactory experience for the visitors. The copy attracted 300,000 additional tourists, strongly indicating the success of this replica (Duval et al. 2019). Some 80% of the visitors to the Chauvet Cave copy felt as if they were in the real cave, not in a replica. This experience holds even if the copy is a scaled-down version of the original. This case suggests that replicated historical sites can adequately imitate real cultural sites; the visitors can experience a new kind of authenticity.

As shown by these examples, tourists who are really interested in art are willing to visit Revived Originals, and they benefit additionally because the paintings there are presented to them in a more attractive environment. Compared to Revived Originals, the access to the original caves is restricted and often completely forbidden. Some critics might argue that this comparison is, therefore, not valid. To counter this argument, it can be pointed out that a lot of people visit the replicated caves of Altamira and Lascaux. This indicates that such newly created cultural supply is accepted. The next example offers a scenario in which both the original and the replica are open and visited simultaneously.

4.3.2 The Tomb of Tutankhamun

Luxor, in the Valley of the Kings in Egypt, contains two tombs decorated with murals of King Tutankhamun from the eighteenth dynasty. One was erected in 1323 B.C. The second is 3 kilometers away and was built in April 2014 as an "exact replica" by the Factum Foundation for Digital Technology in Conservation based in Madrid. The latest 3D methods were used for this project. For an additional entrance fee, both sites can be visited (Wong and Santana Quintero 2019).

In contrast, the original sites of the caves in Altamira and Lascaux can no longer be visited. The Revived Originals are the only opportunity for visitors to see the prehistoric art treasures. Even so, Revived Originals draw attention to these milestones in cultural history and contribute to research and education.

4.3.3 Panoramas and Dioramas

Panoramas offer a wide viewing angle. They are often used to depict spectacular landscapes, for example, the Alps and the skylines of cities. Over time, various methods have been developed for the creation of panoramic images, mainly using 360° paintings and panoramic photography. Dioramas include showcases in which model figures and model landscapes are arranged in front of a painted background in a scene-like manner, following the pattern of Christmas cribs.

The first panoramas date back to the end of the eighteenth century, for example, in Leicester Square in London, where giant paintings were shown until 1861. In other rotundas, spectacular and lifelike scenes were depicted, such as the Battles of Trafalgar and Waterloo.

The panoramas were then developed into large dioramas, in which several hundred people could view the scenes in a rotating hall, in particular by one of the pioneers in photography, Louis Daguerre. Because of their three-dimensional design, they impressed visitors with such intense visual experiences that they could hardly distinguish between paintings and reality (Jung 2020: 51–52).

The *Bourbaki Panorama* dating from 1881 in Lucerne, Switzerland, is one of the few giant circular paintings preserved worldwide. It is 112 meters long and 10 meters high. In front of it is a lively forecourt on which, for example, a railway carriage is placed. This creates a three-dimensional effect. The panorama is a reference to the Franco-Prussian war of 1870–1871 and shows General Bourbaki's French Eastern Army as it took refuge in Switzerland at the end of the war, which France lost. It displays the first humanitarian actions of the Red Cross. All in all, an experience is created, which takes visitors to another place and time.

These panoramas were a sensation at that time and attracted an enormous number of people. They can be seen as early examples of the Revived Originals proposed here. However, today's technical capabilities for creating a virtual reality are far better and result in more exciting representations.

4.3.4 Wittgenstein's Hut in Norway

In the spring of 1914, the eminent philosopher Ludwig Wittgenstein built a house on a steep, tree-covered hillside near the small village of Skjolden on the Sognefjord in Norway, because the hustle and bustle at Trinity College Cambridge disturbed him. He could only reach his house by rowing a boat across Lake Eidsvatnet. In this house, Wittgenstein laid the intellectual foundations for his *Tractatus logico-philosophicus* and *Investigations*. After Wittgenstein's death in 1951, the house was demolished. In 2019, it was rebuilt on the still-existing foundations according to original plans using boards and beams from the original, as far as they were still available (Heim 2019). For Wittgenstein's admirers, the replicated house has become a visitor magnet.

4.3.5 Neuschwanstein Castle

This Bavarian castle, known as a "medieval castle," was built in 1869 and is therefore not medieval at all. Nonetheless, Neuschwanstein is a great tourist attraction.

With up to 6000 visitors per day and 1.5 million visitors per year, the castle is the largest tourist attraction in Bavaria and far beyond. A major reason for the high number of visitors is that the beautiful building really looks like a medieval castle. As can be seen in Fig. 4.2, the castle has impressive features. It certainly looks the way most tourists think a historic castle should.

The Bavarian King Ludwig II spent enormous sums of tax money on the castle and was deposed thereafter for this reason. The close connection of

Fig. 4.2 Neuschwanstein Castle. Source: Picture by Pexels on Pixabay

King Ludwig II with Richard Wagner, who wrote such medieval-inspired operas as *Siegfried* and *Tannhäuser* explains the King's soft spot for medieval architecture, which makes the castle very attractive to visitors.

4.3.6   Berlin and Other German Cities

In Berlin, the city palace, built in 1442 and demolished by the communist regime in 1950, was rebuilt. The reconstruction is limited to the façade, while the building behind it comprises contemporary architecture.

Bombing during the Second World War damaged almost all German cities severely. Some cities were hardly recognizable after the destruction. Two examples are Nuremberg, where 95% of the old town was demolished, and Berlin, where 70% of the buildings were completely obliterated. Both cities have now become major tourist attractions again. The medieval quarters of Frankfurt am Main were also widely destroyed in the war. From 2014 to 2018, some of the houses were rebuilt to their original appearance (Pietersen 2006). These buildings have been approved by the locals and are frequently visited by tourists.

4.3.7   Ballenberg, Open-Air Museum of Switzerland

In this open-air museum, located near Brienz in the Bernese Oberland, over a 100 historic buildings from all parts of Switzerland have been reconstructed in their original appearance. Most of them are farmhouses. Traditional crafts are presented to give a better impression of how peasants did their work in the past. In addition, the affiliated Ballenberg Landscape Theatre performs in this historical setting. In 2008, the museum had around 300,000 visitors, but by 2017 only 200,000, which led to financial problems. The museum is apparently not yet considered as an opportunity to get to know the variety of Swiss house types in one place by foreign tourists.

4.3.8   Swissminiature

This site is located in the Canton of Ticino, in Melide near Lugano. It contains over 130 detailed models of patrician houses, churches, castles, monuments, and other tourist attractions of Switzerland built on a scale of 1:25. There are also 18 miniature trains, as well as several cog railways, cable railways, suspension railways, and models of ships. Switzerland can be explored in time-lapse mode. Visitors will also find a self-service restaurant and a souvenir shop. Around 200,000 people visit Swissminiature every year.

4.3.9 Minimundus in Klagenfurt

This exhibition on the shores of Lake Wörth in Carinthia presents 159 miniature models on a scale of 1:25 of famous buildings, trains, and ships. Minimundus is promoted to visitors as "around the world in one day." One-third of the buildings on display are originally situated in Austria, for instance, the reproduction of the Graz Schlossberg with its clock tower. The nine-level Mayan pyramid El Castillo in Chichen Itza, Mexico, is the centerpiece. Other buildings include St. Peter's Basilica in Italy, Abu Simbel in Egypt, and the Taj Mahal in India. Since its opening in 1985, 15 million visitors have visited.

4.3.10 Replicas in the United States

The Revived Originals proposed here could be seen as an evolution of the various Disneylands. There, some cultural sites have been copied, too, but mostly at a smaller scale than the original. In a hotel in Las Vegas, well-known parts of Venice were copied. The Canale Grande, complete with gondolas and gondolieri, and the Doge's Palace were reproduced on a much smaller size. The situation therefore is totally different from the real Venice, and no effort is made to familiarize the visitors with any historical and cultural context. It is obvious that it is only made to seduce people to spend money in a pleasing environment. The Eiffel Tower, among other buildings, was also replicated in Las Vegas. As Fig. 4.3 shows, the Tower looks impressive within the hotel scenery. Other Eiffel Towers in the United States are situated in Mason, Orlando; Paris, Texas; and Paris, Tennessee; none of them has been built to the same scale.

At the Holy Land Experience in Orlando, Florida, the story of Christ's Passion is presented live five times a week, including flagellation, artificial blood, and crucifixion. Christ is played by an actor (Schader and Aeby 2019).

The fundamental differences from the Revived Original proposed here are obvious. In the Disneylands, the Chinese Window of the World amusement park and other theme parks, and a large number of different sites are situated next to each other. In Macao, an effort was made to improve the replication of Venice over that in Las Vegas. Thus, the Canale Grande and some historic buildings are reproduced in larger size and more carefully. But many

Fig. 4.3 Copy of the Eiffel Tower in Las Vegas. Source: Picture by skeeze on Pixabay

expensive shops, mainly for jewelry and watches, surround these replicas closely. With such an incoherent assembly of sites, visitors cannot experience the context of the culture and history of Venice. It is impossible for them to immerse themselves in its atmosphere as a cultural place. Tourists only gain at most a superficial perception of the lagoon city, and probably associate it more with an expensive shopping experience.

4.3.11   Replicas in Japan

There are several history theme parks in Japan. Edo Wonderland Nikko Edomura is one that represents feudal Japan. It recreates Japanese life during the Edo period, which extends over more than 250 years, from 1603 to 1868. The little town is built in the architectural style of this era and is populated by actors in period costumes. Park visitors can rent a similar costume, engage in games, and see live shows and a variety of theater performances. There are many restaurants and shops, some of which replicate storefronts that show how period craftsmen worked.

4.3.12 Replicas in China

The headquarters of the huge digital company Huawei in Guangdong, China, has attracted a lot of attention. It is located near Shenzhen, a city close to the Hong Kong Special Administrative Region. The group is building a new university for its employees. In an area of 9 km², the company is also erecting various icons of 12 European cities such as Paris, Rome, Bruges, Heidelberg, and Budapest but on a smaller scale than the originals. Figure 4.4 shows the Huawei Base in Songshan Lake District imitating Heidelberg in Germany.

In Shenzhen, an amusement park will copy over 100 of the world's most visited tourist attractions on a smaller scale. In 2012, London's Tower Bridge was copied, with four towers instead of two. And larger ships cannot sail along the canal because the bridge has no lifting mechanism.

In the former Portuguese and now Chinese Macao, the Doge's Palace in the world's largest hotel, the Venetian, has been copied exactly like the original and on the same scale. Additionally, a small part of the Grand Canal has been

Fig. 4.4 Huawei Base in Songshan Lake District imitating Heidelberg in Germany. Source: Huawei Base, China by VogelSP of iStock

reconstructed with gondolas and singing gondoliers. But below the copy of the Rialto Bridge buses and cars pass through, which totally changes the impression, especially because huge hotel buildings dominate these replicas.

Various cities in China have copied the Eiffel Tower, but always at a smaller scale. For example, the Eiffel Tower in Hangzhou, southwest of Shanghai, is 108 m high, which is only one-third the height of the original.

Neuschwanstein Castle was also duplicated in China because it meets the expectations of how a medieval castle looks like in Europe. For this purpose it is irrelevant that Neuschwanstein itself is an imitation and has no historic origin.

In Changde, a provincial city with 6 million inhabitants, a German Quarter has been built, based on the architecture of the Lower Saxony state capital Hanover.

The Titanic is being replicated in its original size in a luxurious hotel complex in Sichuan. The 269-m-long ship will serve as a hotel that lies permanently at anchor and never sails (Joost 2019). Next to it are replicas of various churches of Venice and medieval castles.

The copies in China represent something quite different from the Revived Originals proposed here. They are rarely of original size, are only partially copied, and do not adequately reproduce the original artistic and historical sites. Visitors cannot immerse themselves in the world of the cultural sites depicted but receive only a small and superficial entertainment.

Nevertheless, these imitations in the various parts of the world show the possibilities for creating replicas and prove that such projects attract a large number of visitors and can therefore be financially viable.

5

Revived Originals and Possible Counterarguments

There are, of course, some counterarguments against my proposal of creating Revived Originals. They raise important issues. But even if some of the objections are perfectly legitimate, they should be considered in comparison to the alternatives discussed above. These either cannot be implemented due to conflicts of interest or do not seem to show any productive results. The proposed alternatives only partially take into account current conditions affecting both residents and tourists interested in culture. They may still be subject to overcrowding with long queues, noise, stress, vandalism, crime, and improper behavior. In addition, one should consider the future prospects of overtourism, which paint a gloomy picture. The measures proposed or taken so far do not seem to be an adequate response to this discouraging outlook.

5.1   Original Compared to Revived Original

The main objection to Revived Originals is that culturally interested tourists want to visit the authentic original and not a copy of the cultural site.

The difference between original and copy is the subject of one of the most famous thought experiments in philosophy, the ship of Theseus: Over the years, parts of this ship have to be replaced one after the other. Probably everybody agrees that if only a single plank is replaced, it remains still the original ship. Although it has changed slightly, it is still the ship of Theseus. Does this still apply if over a longer period ever more parts are replaced until the ship finally consists completely of new material? Philosophers still discuss this question intensively.

© The Author(s), under exclusive license to Springer Nature Switzerland AG 2021
B. S. Frey, *Overcoming Overtourism*, https://doi.org/10.1007/978-3-030-63814-6_5

The assumption that tourists who are interested in culture only want to visit the original and not a replica can be refuted with several arguments.

In many cases, visitors cannot distinguish the original condition from the newer additions. An example of this is the Campanile di San Marco (St. Mark's Campanile) in Venice. It was erected in the years 888–911 and was repeatedly modified until 1517. The tower then collapsed in 1902 but was rebuilt exactly as it was afterwards. So, the tower existing today dates back to the year 1912, neither 911 nor 1517, but surely only a few tourists are aware of this. Even if they were, it is unlikely that they care. Even for people valuing authenticity, replicas of some parts of a cultural site are perfectly acceptable. Actually, St. Mark's Square is not affected by the fact that the Campanile is a copy at all.

Another noteworthy point is that in Venice many restorations in the late nineteenth century followed a preservation approach by replacing deteriorated parts. A prominent representative of this philosophy was Giovanni Battista Meduna, who also contributed to the restoration of the San Marco tower in the role of the chief restorer. Using the prefabrication method, which is criticized for being imprecise, he replaced many deteriorated elements, even modifying the color intensity. Thus, the historical mosaics and the original message were even lost before the tower collapsed, but this fact is rarely acknowledged or known by most visitors (Basilica di San Marco 2020).

For cultural sites, especially historical cities, it is impossible to maintain the original condition. Old buildings decay or are demolished and then new buildings are erected. Thus, the original city changes continuously. So, depending on the moment at which it is visited, people experience a different original. Changes also affect other kinds of sites and even specific artworks.

The original tomb of Tutankhamun was covered with a thin layer of dust over time. The recent replica of the tomb in Luxor imitates this dusty patina. The original tomb, on the other hand, has been carefully cleaned and now has a different appearance than before (Wong and Santana Quintero 2019). It is unclear which tomb should be considered being the "real" one.

There is a traveling exhibition titled "Tutanchamun: His Tomb and the Treasures." A profit-oriented company presents exact copies of the tomb and of the additional objects contained in it. The exhibition has been shown in many countries. The entry price is quite high. In Zurich, for instance, an adult had to pay 28 Swiss Francs, which is expensive compared with other local museums such as the Kunsthaus (16 Francs) or the Swiss National Museum (10 Francs). Nevertheless, more than 6.5 million people worldwide have visited the traveling exhibition so far. The show does not present any original pieces of Tutankhamun's tomb, but perfect replicas. They look as they did in the original state when the pharaoh was buried more than 3,400 years ago. The replicas offer viewers a superior picture than when the Tomb was discovered and explored by Howard Carter in 1922. The organizers claim that the visitors can see the treasures better than "museum tourists in Egypt." The fact that Tutankhamun's mummy can only be visited in its tomb in the Valley of Kings and not in the Egyptian Museum in Cairo supports this statement. The exhibition does not use any digital means to make it more attractive. The huge success of this exhibition, which consists solely of copies, suggests that people enjoy seeing them if they are well presented.

Which condition should now be considered the "original" one? This question is to a great extent irrelevant (see extensively Frey 2003, Chapter 12). An original is always a product of its time. Can Mozart's oeuvre only be performed authentically on a genuine old instrument of his time? Or is it allowed to play his music with modern instruments that are in some ways superior to the traditional ones? All the aspects mentioned above make it difficult to draw a clear line between copies and originals. The Revived Originals, I propose, can document the condition at different periods of time, which is a great advantage. For instance, the city of Venice can be presented in different epochs with the help of virtual techniques (which is also the goal of the Venice Time Machine project, cf. Hafner 2019). Thus, visitors to a Revived Original can experience how a cultural site changes over time, which provides additional benefits and facilitates historical understanding.

One criticism of the glorification of originals goes even further. It is argued that the Disneylands, which replicate certain objects in the United States (e.g., Wild West streets), are more real than the "real" sites (Baudrillard 1994): they represent America as it "really" is. This argument could also apply to many other art sites.

Copies are also often used to protect originals from destruction so that the heritage of mankind can be preserved. For example, copies replaced the English crown jewels after the IRA attacked the Tower of London in July 1974.

Many antique copies are highly valued today. The marvelous sculpture of *Aphrodite or Venus of Knidos* in the Capitoline Museums is not the original by Praxiteles from 350–340 BC, but a Roman copy which was made much later. The same applies to the well-known *Discus Thrower* by Myron in the Museo Nazionale Romano, which also dates back to Roman times. The world-famous painting of the *Last Supper* in the Church of Santa Maria Delle Grazie in Milan has been restored and painted over so often that it is almost impossible to recognize the hand of Leonardo da Vinci in it. But the painting still attracts a great number of visitors who obviously enjoy the experience. In the Renaissance, the young Michelangelo copied a sculpture by his teacher

Domenico Ghirlandaio. This work was not considered a fake or of lesser quality, but a breakthrough that helped Michelangelo to become one of the greatest artists of all time. Few people would argue that his copy is of little esthetic or cultural value. On the contrary: most viewers consider it one of the most beautiful pieces of art produced by a genius.

None of the cultural cities and sites mentioned so far are original in the sense that they have not changed over time. Many cultural monuments have collapsed or been destroyed by external influences. Modern artists such as René Magritte or Salvador Dalí deliberately destroyed the difference between original and copy as part of a revolt against the burden of the past.

My proposal of Revived Originals uses copies for another purpose: to introduce the cultural treasures of humanity to tourists who would otherwise be less informed and scarcely involved. To understand art properly, it is key to understand the changes taking place over time. Revived Originals can reproduce the cultural sites in various periods, and can therewith help viewers to better understand the development of art sites in the course of history.

Another argument is that tourists who are less interested in culture just prefer to shoot their selfies at the original sites. They want to demonstrate to their friends and followers that they were actually there. So they visit the most famous destinations and take the same pictures as have been taken millions of times before. The so-called St. Matthew effect ("For to everyone who has will be given, and he will have more") takes place: Sites, which were already crowded, become even more crowded.

But if the places are already famous and appreciated, friends and followers can also be impressed by a visit of Revived Originals. After all, many people post their selfies from the various scenes in the Europapark or Macao, although these are not the original places. Posts from newly created and unusual cultural sites may become even more desirable.

Cultural sites that become world-famous through films, television series, and video-clips can be particularly attractive to tourists. These people only know the locations from their TV or mobile and may therefore prefer replicas that are easily accessible and presented in a very visitor-friendly manner. For them, it makes little difference whether they visit, for example, the film sets from the series *Game of Thrones* in Dubrovnik or a Revived Original of the city in its vicinity. The same applies to the small town Hallstatt, which became widely known through the movie *Frozen*. Many tourists, also those interested in culture, like to visit film studios where famous films were shot. This is not only true for studios in Hollywood. Many tourists, for example, visit the Moroccan city of Ouarzazate, because parts of the films *Lawrence of Arabia, Gladiator, Kundun,* and *Asterix and Obelix: Mission Cleopatra* were shot there.

It would be difficult to see a difference between the film set and the original building, as shown in Fig. 5.1. The replica meets the expectations and requirements of most tourists intrigued by buildings in ancient Egypt.

5.2   Missing Atmosphere and Acceptance

Culturally interested tourists in particular attach great importance to the ambience of the sites they want to visit. A major aspect is the local life. For many tourists, experiencing the authentic life that the residents of a site lead is key.

Today, in the age of mass tourism, the desired local color no longer exists in many places. In the cultural sites, which suffer most of overtourism there is little or no local life remaining. Tourists flock to places such as Dubrovnik, Riquewihr and Rothenburg ob der Tauber but no longer encounter locals. They do not meet them in the bars or restaurants (because they are too expensive for the locals) nor do they see children playing in the streets or young local people having fun together. The only exceptions are the few residents

Fig. 5.1 Film set in Ouarzazate in Morocco. Source: Picture by Bruno S. Frey

working in the tourism sector. However, they have other tasks than to chat with tourists. So tourists only meet other tourists (D'Eramo 2018), which makes it more likely that they talk to another visitor rather than to a local.

A Revived Original can, at least partially, reestablish the original authenticity of an art site. Thus, the Revived Originals should be situated as close as possible to the original sites. For instance, the reproductions of the caves of Altamira and Lascaux with their prehistoric paintings were built just a few hundred meters away from the original caves. The replica of the tomb of Tutankhamun in the Valley of the Kings is also only 3 kilometers away from the original. All external factors such as landscape, climate, weather, and the local language are thus identical to the surroundings of the original, which makes the Revived Original appear more authentic to visitors.

But more importantly, virtual technology allows tourists to become involved and to immerse themselves in the culture and history of a site. This provides an additional experience for visitors. They enter a different world—in contrast to the many ignorant tourists who visit an art site but have little or no idea of its history or cultural significance.

When the Revived Originals are a success, the original sites will become less crowded. Locals who still live there will be able to live a more normal life again. At the same time, this will make the ambience more pleasant for culturally interested tourists who particularly value the special atmosphere of a place.

Residents living close to cultural sites may have strong ownership feelings from their sense of place. This drives the focus on community conservation and results in a backlash against tourism. Although local residents may benefit from tourism, they are not entirely dependent on tourist income, which results in tourism activities being seen in a more negative light (Amore 2019; Ashworth and Page 2011; Rasoolimanesh et al. 2017).

The goal is to achieve a healthy level of tourism, still generating income and jobs for the locals but not harming the quality of life or cultural heritage.

5.3   Not Interesting for Selfie Tourists?

As already mentioned, some visitors are not genuinely interested in the cultural sites. They just want to show off to their friends and followers on social media to have visited a famous place. They post selfies to document their visit. However, the culture and history of the site may be only of little interest to them.

This group includes many Asian tourists, especially those from China, but also increasingly young visitors from Europe who have grown up in the digital world: the "digital natives." This group of people, which is a growing percentage of all tourists today, tends to consider a visit to a cultural site as done, as "ticked off," when a selfie has been made ("been there, done that") (Neuhaus 2019). Influencers also play a significant role. They show certain cultural sites to their followers and encourage them to visit these places and to bring it to the attention of their families and friends by taking selfies.

At the original location, it is often difficult to take selfies in such a way that viewers can immediately recognize the location. As a result, the right position in front of a famous and iconic site is crucial in the internet world. However, people with such intentions often struggle to find a suitable position for their selfie, because such a spot is often particularly crowded. So they hustle for a good position, where they are in the best place in front of the desired spot. This phenomenon can be observed at the Norwegian Trolltunga. Instagram has led to an incredible increase in tourists there, too, from 1,000 to 100,000 in just 5 years. Nowadays, tourists have to be prepared for a waiting time of 60–90 minutes (Pitrelli 2019). But the pictures shared on Instagram, of course, do not show the long queues because this would damage the impression of uniqueness.

It is easier for replicated cultural sites to implement and provide this function than for an original site. A Revived Original can be built in a way that ensures selfie shooting is easy and effective, and selfie spots are marked on the ground. What matters is, that the post about the visit gets many likes from followers. This mainly depends on the quality of the Revived Original, such that the followers may not even be able to tell the difference from the original site at all.

Furthermore, the dreaded "No photos" signs that can be found at many special cultural sites can be replaced by "Please take as many photos as you want." This would open completely new opportunities for selfie enthusiasts increasing the glamour factor of Revived Originals. The "Do not touch" sign may even be substituted by an invitation to feel with one's hands the reproductions shown in Revived Originals. This opportunity would be welcomed not only by people who are blind or have poor vision but also by many other people who put a special value on haptic experiences.

6

What Problems Are Likely to Arise?

Successfully creating Revived Originals requires considerable financial resources, but above all, it requires initiative, assertiveness, originality, and a good sense of design. Accordingly, interdisciplinary cooperation is necessary. An example is the replica of the tomb of Tutankhamun in Luxor. At least seven different professions were involved in its construction, including archaeologists, environmental engineers, microbiologists, digital recording specialists, architects, designers, and conservators.

6.1 Which Organizational Forms Are Reasonable?

Private entrepreneurs can create Revived Originals if they can raise the necessary financial means. The funds needed depend mainly on the size of a Revived Original. If parts of a city such as Venice are copied and placed at another location, and virtual techniques are used, a great many resources are needed. Excellent entrepreneurial skills are critical. If, on the other hand, individual cultural sites or even only individual cultural monuments are replicated, less effort is required.

In many cases, a public–private partnership with the authority of the state may be optimal. The wide range of skills of employees in public institutions, especially theaters and museums, can be very useful. The experience gained through scenography, which today is used in many such locations, is very helpful to create Revived Originals.

© The Author(s), under exclusive license to Springer Nature Switzerland AG 2021
B. S. Frey, *Overcoming Overtourism*, https://doi.org/10.1007/978-3-030-63814-6_6

Cooperation with the state, especially at local and regional levels, also makes it easier to deal with administrative obstacles that are otherwise difficult to overcome. The creation of Revived Originals may also face copyright problems, which can be solved at least partially through a partnership with the government. However, in most countries, the restrictions on rebuilding historic monuments expire 100 years after their construction.

6.2   Is Suitable Property Available?

The building of a Revived Original requires an area of free and appropriate land. For example, if Venice were to be replicated, an island or a shore of a sea or a large lake would be suitable. At the same time, Revived Originals must be easily accessible for tourists; therefore, good public transport connections are particularly important.

In principle, Revived Originals can be placed anywhere. Not only sites close to the original locations can be considered, as is the case with Altamira and Lascaux, but they can also be in another country or even on another continent. However, it is favorable to build the Revived Originals in a location near the original. This can better evoke the specific atmosphere of a cultural site. It is certainly less authentic if you locate a new Venice somewhere in Australia or India than in the immediate vicinity of the original.

Producing identical copies for Revived Originals may be too complicated or even impossible. However, today's 3D technology allows us to replicate artworks and buildings far better than before, so that hardly any difference can be noticed. In some cases, the options for Revived Originals are limited, as the surrounding landscape also plays an important role. But nowadays even a landscape can easily be represented artificially by virtual techniques.

There are many intermediate solutions between identical copies of the buildings of a cultural site and purely virtual replica. For example, a part of a cultural site can be reproduced identically and another part virtually. For visitors, transitions between these two forms are often hardly noticeable. The more expensive an identical replication of buildings is expected to be, the more likely it is that a virtual reconstruction will be considered. The ratio thus depends on the relative costs of the two options.

6.3   Is There a Rivalry with Existing Replicas?

It might be objected that the copies of cultural sites already built in China and elsewhere make it unnecessary to visit the Revived Originals proposed here in the vicinity of the original cultural site.

There are two counterarguments against this view.

First, the existing copies are technically less advanced than the Revived Originals. For example, the Venice presented in a hotel in Las Vegas has a much smaller size and is, moreover, not carefully crafted. These shortcomings are immediately noticeable to every visitor. The reconstructed Grand Canal in Las Vegas is not only small but also not similar to the original at all.

In Macao, a part of Venice is copied in the world's largest hotel, the Venetian. The front of the Doge's Palace, which is reproduced in its entirety and with great precision, serves as the hotel's facade. Figure 6.1 shows that the Palace really looks like the original in Venice. A lot of gambling rooms are directly behind the Doge's Palace.

The Rialto Bridge is also reproduced, but, in Macao, it arches over a street filled with cars and buses instead of a canal. The Canale Grande is represented

Fig. 6.1   Copy of the Doge's Palace in Macao. Source: Picture by Bruno S. Frey

as a short rivulet with singing gondoliers rowing on it. Even though no visitor from the Chinese mainland believes that they are seeing and experiencing the real Venice, there are still long queues of tourists eager to board the gondolas.

Slightly more similar to the Revived Originals proposed here is the small Austrian town Hallstatt, which is partly copied in Guangdong, a province in South China. The town center with copies of the church, the fountain, and other buildings is built on a lake. Nevertheless, it is obviously different from the original, not least because the atmosphere of a little Austrian village is missing.

Second, the existing copies of cultural sites are not, or only insufficiently, supplemented with digital technology such as augmented and virtual reality. At best, visitors gain only a faint impression of the culture and history of a place. The Revived Originals proposed here are considerably different: visitors are able to immerse themselves in the cultural site and thus enjoy an experience that cannot be achieved only by copying some buildings and other features.

6.4   Will the Originals Still Risk Being Overcrowded?

A Revived Original could encourage tourists to visit the original cultural site later. In this case, the replica and the original would not be substitutes but would actually complement each other. Cultural overtourism might even be exacerbated. Some tourists will certainly visit the original in the future, which can be considered a positive effect: a new category of people will be culturally stimulated. Revived Originals can awaken a desire to experience the original place in people who have been less interested in the culture so far.

However, a greater rush to the original art sites is not to be expected, for several reasons. Because some tourists from other continents who have visited a Revived Original will only travel to Europe once, they will not have the opportunity to see the original site. Many tourists will be satisfied with the Revived Original and therefore do not care to see the original cultural site in the future. Instead, they will visit other cultural sites, of which there are many in the world. In addition, a reverse effect is also conceivable. Those who have seen the original may also consider visiting the Revived Originals with the exciting virtual technology. Those people who liked the Revived Original better than the original cultural site will also visit replicas of other cultural sites in the future. As a result, the original sites will be less crowded and cultural overtourism will be curbed.

6.5   Political Resistance

Not all individuals and social groups will welcome the idea of Revived Originals, despite their great advantages. Instead, strong resistance is to be expected from some groups.

Local commercial service providers are threatened with losses in sales and profits, which is why they will take action to defend themselves. Those affected are hotels and restaurants, souvenir shops and other shops visited by tourists, apartment landlords (especially local owners of apartments which are rent out at short notice, e.g., via Airbnb, and which generate high revenues) and parking lot managers. For example, owners of souvenir shops will not want to lose visitors, even if this reduces noise and improves the cleanliness in the place. The decisive factor is whether the other commercial service providers would also agree on limiting the number of tourists. Even if they would do so by taking collective action, the question remains whether, overall, the improvements resulting from a better quality of life due to less noise and greater cleanliness would outweigh the loss in profits. It should be noted that many commercial operators are not based in the town and therefore pay little regard to the quality of life of local residents.

Resistance to Revived Originals can also be expected from the employees in the cultural sites. They fear losing their jobs.

Environmentalists may fight against Revived Originals if they occupy natural, previously vacant land. However, these concerns must be compared to the fact that the newly created Revived Originals are particularly eco-friendly; visitors cause less environmental damage, especially in the long term. In contrast, original sites are ecologically unfavorable in many respects because they were built in past centuries. This applies not only to the environmental damage caused by heating in winter and cooling in summer, but also to the collection and purification of wastewater, and other environmental threats. In this respect, some art sites are in a precarious state; Revived Originals represent a major step forward in comparison.

Some defenders of cultural sites, such as those on the World Heritage List, will also be opposed to Revived Originals. Reconstruction will be strongly rejected because the historical sites are regarded as unique. To partially rebuild them elsewhere and enrich them with augmented and virtual reality technology is considered blasphemy.

This argument ignores the fact that it is wrong to take an existing art site as unalterable. Paris provides a good example. When the Eiffel Tower was built to commemorate the 100th anniversary of the French Revolution on the occasion of the World's Fair of 1889, there was strong resistance from the intellectual elite to this building, who found it disgusting. Numerous personalities, including Charles Gounod, Alexandre Dumas, Charles Garnier, and Guy de Maupassant, were outspoken critics. They considered the tower unnecessary, ridiculous, and monstrous. As passionate lovers of the hitherto untouched beauty of Paris, they protested in the name of well-established French taste (see Muscheler 2008). The ostentatious iron construction of the Tower without a façade was deemed to be scandalous. Today, with around seven million visitors each year, the Eiffel Tower is one of the most famous landmarks in the world. It is regarded as a masterpiece of architecture and engineering. For many visitors, it represents the essence of Paris. Figure 6.2 shows the long queue of tourists wanting to enter the Tower.

Revived Originals could also be rejected because they could create a two-class tourism: the well-educated and wealthy tourists will visit the original site; the less educated and poorer ones will have to be content with the Revived

Fig. 6.2 Queue to visit the Eiffel Tower. Source: Picture by PublicDomainPictures on Pixabay

Originals. However, this fear is not justified. Revived Originals offer additional services that are largely lacking in the original places. In particular, they employ augmented and virtual reality technology to give visitors completely new and fuller impressions of history and culture. Tourists are likely to accept these additional services. They will also enjoy Revived Originals as they are more easily accessible and provide more amenities. These advantages are not only important for families with children but also for tourists who appreciate new and creative concepts. Experience with the Revived Originals of the French and Spanish caves (Lascaux, Altamira) and the tomb of Tutanchamun support this view.

7

How Can Revived Originals Become Reality?

The concept of Revived Originals can be applied to diverse cultural sites. Cities, towns, and individual artworks can be copied and linked to culture and history using digital technologies.

As an example, a proposal is first made on how Venice, one of the cities, which suffers most from overtourism, could be replicated. Then we consider how this could be applied to six North Italian and four Austrian cities, which are particularly overcrowded.

These suggestions should be considered as preliminary ideas and in no way claim to be directly applicable. Rather, they are intended to show readers how Revived Originals could be designed in practice.

7.1 Venice

A Revived Original of Venice will certainly include the Doge's Palace, St. Mark's Cathedral, the Campanile and parts of St. Mark's Square, and the Rialto Bridge. Other parts of the city, including sections of the Grand Canal with its palaces, can be reproduced realistically using augmented reality. Ideally, visitors will not be able to recognize which part of a building or district has been replicated in masonry or is only virtually represented (Losse 2019).

Historical figures from various periods in the history of Venice can be vividly represented by holograms in the Revived Original. They relate to politics, music, paintings, theater, the church, and particular historical episodes.

B. S. Frey, *Overcoming Overtourism*, https://doi.org/10.1007/978-3-030-63814-6_7

- The sophisticated way of electing the Doges of Venice by both voting and random selection can be shown. This includes identifying the various families and other groups dominating the city government, as well as those parts of the inhabitants of Venice that were excluded from political decision-making. The reasons why the Doges were partly elected by lot, instead of by direct voting, is also likely to be of great interest to visitors.

- The life of prominent Doges can be demonstrated: The families they came from, how they were elected, what they achieved, and how their lives ended. An example could be the 77th Doge of the Republic of Venice, Andrea Gritti. Two great artists painted this politician. Tintoretto's paintings are in the Doge's Palace and Titian's in the Metropolitan Museum in New York and the National Gallery in Washington. Gritti could be presented by a hologram at his political work in the Doge's Palace, the people he conversed with, and the way he ruled.

- Debates in the various public bodies of the Serenissima—as Venice is also called—could be transformed virtually. The visitors can be informed about the political life in the city as well as the economic activities undertaken, including the extensive trade relationships in the Mediterranean Sea. The conflicts that existed between various groups in the city can be pointed out.

- The music of famous composers, such as Antonio Vivaldi, Tomaso Albinoni, and Claudio Monteverdi, could be brought to the visitors by presenting excerpts or entire works performed visually in a relaxed atmosphere. Furthermore, the lives of these artists can be explored.

- Painters such as Bellini, Titian, Tintoretto, Tiepolo, Guardi, and Canaletto can also be followed as holograms while painting their masterpieces. The paintings can be brought to life by showing their development and how they look now. The origins and achievements of the persons painted can be illustrated by virtual reality and holograms.

- The works of the comedy poet Carlo Goldoni can be revived. By digital technology the background and content of his works can be brought to the attention of the tourists. Moreover, the effect on the contemporary society as well as the relevance for today can be explored.

- No fewer than eight popes have come from Venice; the last, but only very briefly, being John Paul I. Their lives can be depicted in a Revived Original. The period in which they lived, and the religious and political challenges with which they were confronted can be explained to the visitors.

- The experiences of the world traveler Marco Polo, who inspired Europeans with reports of his journeys through China, can be made known. Such information is especially interesting for today's visitors from that country.

– The unique escape of Giacomo Casanova from the lead chambers of the prison in the Doge's Palace can be depicted in an exciting way. A hologram could be designed which would make these and other events in his life even more attractive to tourists.

– Historical events can be dramatically illustrated, such as the unfortunate end of the Republic of Venice in 1797 by Napoleon Bonaparte after 1,100 years of its existence. The visitors of a Revived Venice can be informed about the circumstances and the background that led Napoleon to his decision.

Visitors to the replicated Venice will be well aware that they are not in the original place. But intense historical and artistic events presented by digital technology will help to overcome this feeling. In the Revived Original, the tourists can gain an idea of what Venice might have looked like in the past, especially when the city was at its cultural peak in the seventeenth and eighteenth centuries during the Renaissance and Baroque periods.

The prerequisite for such valuable experience is a sensitive yet easily accessible and exciting presentation in a Revived Original. It is scarcely an exaggeration to suggest that the replica Venice will seem more real than the original, because there the cultural and historical situation and development of this city is convincingly conveyed to the visitors—which is rarely the case with visitors to the original city.

7.2 Combining Several Cities

It is known that many tourists can afford only little time to stay in one city or region. It has been reported that for many tourists, 1–2 days must suffice to visit the whole of Northern Italy (García-Palomares et al. 2015; Ram and Hall 2017). This is also evident in Hallstatt, where the average stay has fallen from 2.7 days in 1990 to 1.4 days in 2016 (Bundesanstalt Statistik Österreich 2019). It may therefore be reasonable to collect the cultural sites of various cities in one place and to use digital technology to create an exciting experience. Disneyland and other visitor parks have been using a similar strategy for a long time by stringing together the most diverse places next to each other. However, this often results in a hodgepodge of various art sites, which can be seen more as an amusing illustration than as serious historical and cultural replicas.

Meaningful Revived Originals of several cities must look quite different from a replica of a single city. Two combinations of such cultural sites are discussed here: North Italian and Austrian cities.

7.2.1 North Italian Cities

There are a large number of cities in Northern Italy that are culturally remarkable. At this point, only one possible selection will be listed. Many other cities could be considered. In addition to Venice, Florence is excluded from the very outset because it is an obvious choice. Both cities certainly deserve their own Revived Originals.

– *Verona*

The Amphitheatre built by Tiberius in 20 AD dominates this city. Figure 7.1 shows this imposing building. It was erected half a century before the Roman Colosseum. This arena is particularly famous because every year a much-visited opera festival is held there. The spectacular performance of Verdi's *Aida* attracts particular attention.

Verona is also well known to many visitors because it is the site of Shakespeare's *Romeo and Juliet*. At the Palazzo Casa di Julietta in the old town, there is a balcony that appears in the drama. However, the balcony was not added until around 1930 and has no authentic connection to Shakespeare's play. Nevertheless, it is a big tourist attraction. Something similar happened with the triumphal arch Arco del Gavi. It dates back to

Fig. 7.1 The Roman Arena in Verona. Source: Picture by Gianni Crestani on Pixabay

the first century AD but was demolished by the French in 1805 and only was rebuilt in 1932.

If Verona becomes part of the Revived Original of North Italian cities, Romeo and Juliet's balcony and the Arco del Gavi could be reproduced without hesitation as they are copies themselves.

Surely a considerable effort is necessary for the replication of the arena. To gain a convincing impression of Verona, a large part of the arena could be presented with augmented reality. This would also better suit the expectations of many tourists.

– *Siena*

The central square, Piazza del Campo, is not only impressive in itself but also famous for its spectacular horse races, held there twice a year. These competitions, in which 17 districts (*contrade*) compete, have been held since the Middle Ages. Figure 7.2 shows this scene and some of the many spectators enjoying the spectacle. In this square are also the Palazzo Pubblico, begun in 1297, where the city government resides, and the Torre della Mangia, which can be seen from far away.

A Revived Original could realistically replicate the Palazzo Pubblico, the Torre della Mangia, and perhaps parts of the Piazza del Campo.

Fig. 7.2 Horse Race on the Piazza del Campo in Siena. Source: Picture by Anastasia Borisova on Pixabay

Digital techniques allow reproducing the Piazza as a whole and would enable visitors to feel involved in a horse race. This would allow them to immerse themselves in an exciting way into the most important event in Siena.

— *Pisa*

The Leaning Tower in Piazza dei Miracoli, next to the medieval Cathedral of Santa Maria Assunta with the Baptistery, is known worldwide. Figure 7.3 shows this extraordinary Campanile. With modern construction techniques, this Campanile can easily be historically replicated. The surrounding buildings can be presented virtually.

— *Padua*

The most famous monument is the important sanctuary Basilica di Sant'Antonio, which contains the tomb of a saint and a high altar with bronze statues by Donatello. The Scrovegni Chapel with its cycle of frescoes by Giotto is particularly important in art history.

In the Scuola di Sant'Antonio, built in 1427 as a guild building, Titian decorated the chapter house with frescoes, such as the miracle of St. Anthony. This vivid painting is shown in Fig. 7.4. This painting can well be replicated as part of a rebuilt chapter house. This offers a welcome opportunity to show visitors the life and work of Titian, including his relationship with other artists and with the principals hiring him to engage in painting particular scenes.

— *Bergamo*

In the upper town, surrounded by massive walls, is Piazza Vecchia with a medieval Palazzo della Ragione (Town Hall). The Cattedrale di Sant'Alessandro Martire and the Capella Colleoni are located in the Piazza Duomo. This wonderful city is illustrated in Fig. 7.5. Tiepolo painted various frescoes in these churches. The opera composer Gaetano Donizetti was born in Bergamo.

A Revived Original could recreate these churches and, using virtual techniques, fill Piazza Duomo with life. In this way, the life and work of the painter Tiepolo can be depicted in an engaging way using these churches. Donizetti's operas of the bel canto of the first half of the nineteenth century, such as *Anna Bolena* (1830), *L'elisir d'amore* (1832), and *Don Pasquale* (1843), now part of the standard repertoire of opera houses worldwide, can be presented to visitors in parts, thus making a visit to Bergamo even more lively.

— *Vicenza*

This city is closely associated with the architect Andrea Palladio. This Renaissance personality built the first freestanding post-antique theater, the Teatro Olimpico, in the city center.

Fig. 7.3 The Leaning Tower of Pisa. Source: Picture by Bruno S. Frey

Fig. 7.4 Giotto: Judas' Kiss and Capture in the Scrovegni Chapel. Source: Picture by Valentina Ficuciello on Pixabay

Fig. 7.5 Bergamo Alta. Source: Picture by nono_08450 on Pixabay

Fig. 7.6 The Villa Rotonda or Villa Capra near Vicenza. Source: Picture by Flavio Vallenari on iStock

Palladio is also well known for having designed many villas in the region of Veneto. The most famous is the Villa Rotonda (or Villa Capra) a little outside of Vicenza. This impressive Villa is pictured in Fig. 7.6. The magnificent buildings designed by this great architect could be copied in part or as a whole, and virtual techniques could show the lives of the inhabitants in various historical periods. With the help of digital techniques, the impact of these buildings on the later architecture can be illustrated.

The six cultural cities of Northern Italy mentioned above—Verona, Siena, Pisa, Padua, Bergamo, and Vicenza—have all been included in UNESCO's World Heritage List. They represent only a small fraction of the many important cultural sites in this region. There is also a wide range of other possibilities. For example, the magnificent mosaics of Ravenna could be included, or the cities Mantua, Sabbioneta, Ferrara, and Modena, which also have important cultural monuments. In addition to Venice and Florence, other large cities such as Parma, Milan, Turin, and Genoa also have many culturally valuable sites worthy of a Revived Original. In all these cases, the goal is not to simply reconstruct some buildings but rather to fill them with life using the most advanced digital technology.

7.2.2   Austrian Cities

Similar to Northern Italy, the most important sights of a selection of Austrian cultural cities can be collected in one place. The selection made in the following can only serve as an example; many other possibilities are conceivable and useful.

– *Innsbruck*

   The icon of this city is the late gothic Goldenes Dachl at the Neuen Hof. Also much visited is the Hofkirche with the tomb of Emperor Maximilian I. Impressive larger-than-life bronze statues surround the mausoleum. These art sites can be used in a Revived Original to illustrate to visitors the rise of the Tyrolean Habsburg dynasty to emperorship.

– *Salzburg*

   Most tourists associate Salzburg closely with Wolfgang Amadeus Mozart. The birthplace of this brilliant musician in Getreidegasse is a major tourist attraction in the old town. It is shown in Fig. 7.7, which reveals its medieval character.

   The city is also known for its music festival, founded in 1920, which is now complemented by the Easter and Whitsun festivals. German-speaking visitors are also familiar with the annual play *Jedermann* by Hugo von Hofmannsthal.

   A Revived Original could rebuild parts of the city, but above all, it could present the life and work of Mozart in such a way that visitors otherwise not much interested in classical music will also find it enjoyable. The visitors could follow Mozart's beginnings in the context of his family, most importantly his father and sister, his many performances all over Europe, his life in Vienna, as well as his early death.

– *Vienna*

   Almost all tourists of the Austrian capital visit St. Stephan's Cathedral, the Hofburg with the Spanish Riding School, the Belvedere Palace, and the Vienna State Opera. Also, the summer residence of the emperors in Schönbrunn attracts many visitors. Figure 7.8 provides an impression of this huge palace.

   Vienna is a city of music. Great artists such as Ludwig van Beethoven, Wolfgang Amadeus Mozart, Johannes Brahms, Richard Strauss, Alban Berg, and Arnold Schönberg have lived and worked there. The Viennese waltz was influenced above all by the Johann Strauss dynasty.

Fig. 7.7 Getreidegasse in Salzburg. Source: Picture by Hans Braxmeier on Pixabay

Painters are also important in Vienna. These are above all Gustav Klimt, Egon Schiele, and Oskar Kokoschka from the twentieth century. Their main works of art can be presented, and in which way these artists influenced each other and how their lives developed.

There are many worthwhile opportunities for a Revived Original. It would probably be too costly to rebuild, for example, the huge Schönbrunn Palace. But with parts of it completed by augmented reality, the visitors could get a good impression.

– *Hallstatt*

This small town of 750 inhabitants in the Salzkammergut region (see Fig. 3.3, p. 38) is visited by tens of thousands of tourists every day. This large number seriously transforms the habitat and quality of life of the local population. The number of visiting coaches, especially with Chinese tourists,

Fig. 7.8 Castle Schönbrunn near Vienna. Source: Picture by Federlos Blog on Pixabay

has increased from 10,000 in 2015 to almost 20,000 in 2018 (Benz 2019). Hallstatt is considered so attractive, especially by Asian travelers, that a partial replica has been rebuilt in Southern China. Many other small settlements in Austria could just as well be incorporated into a Revived Original.

7.3 What Is Required to Create Revived Originals?

A bundling of different cultural sites in one place, as outlined for Northern Italian and Austrian cities, requires considerable effort. Experts from different disciplines must work closely together.

- *Art*

  Suitable cultural sites have to be chosen. The UNESCO list of World Heritage Sites is too long; it contains about 900 particularly important cultural sites. Therefore, a selection is necessary. Special attention must be paid to the question of which cultural sites are threatened. Reasons include wear and tear by too many tourists; damage by environmental impairment, especially air pollution; and possible destruction by terrorist acts but above all also by unwise political measures in which culturally valuable sites are sacrificed in favor of modern buildings.

— *Architecture and urban planning*
Which cultural sites can be copied? Are there specialists, especially crafts-people who are capable of doing this? What are the costs involved? Can the replicated cultural sites be sensitively placed next to each other, even if they belong to different styles, such as Romanesque, Gothic, Renaissance, and Baroque? How can the transition to modern buildings, which are necessary in the Revived Originals, be managed?

— *Tourism*
What is the target audience of a Revived Original? Do they particularly attract tourists interested in art or are they also attractive to other visitors? What distribution of the tourist flow between the original and the Revived Originals can be expected, depending on entry prices? How can an atmo-sphere be created in the Revived Originals that will find acceptance? What amenities for visitors, including cafés, restaurants, hotels, and souvenir shops, must be provided? And this is not to forget the planning of access by public transport as well as by private cars and buses and the essential parking lots.

— *Ecology*
How can Revived Originals be designed in an environmentally friendly way? Not only can the construction result in zero-carbon-emission build-ings, but the logistics of arrivals and departures should also be planned to reduce the burden for the environment. Environmental and sustainable implementation will promote their political and social acceptance.

— *Economy*
The creation of Revived Originals requires a careful investment calculation. The future expected costs have to be compared to the expected revenues. The commercial return on a Revived Original depends on the number of visitors and the entry fee. An appropriate entry price must be carefully chosen. It must be considered how potential visitor groups react to various admission prices. This depends in particular on the alternatives available to them. For example, if the original venue does not charge any, or only a low, entrance fee, the admission price for its Revived Original must also be kept low. However, tourists are quite used to paying rather hefty admission fees for leisure facilities such as Disneylands and Europapark Rust. For a family, the amount can easily exceed 100 euros. Both the number of visitors and the possible entrance fees can only be approximately estimated. Therefore, evaluating future revenues demands consideration of a range of scenarios. The costs for the construction of the replicated buildings and comprehen-

sive digital design can probably be calculated more easily. The same applies to the costs of personnel, maintenance, and insurance. But even here, many things remain uncertain.

– *Finance*

Creating a Revived Original requires high financial expenditure. The return to be gained must be compared to alternative investment opportunities. Whether the banking system is willing to provide the necessary loans is by no means certain. State support is also conceivable because the Revived Originals promote the preservation of the original cultural sites, which means a service is provided for society as a whole.

This list makes clear the high demands placed on Revived Originals and their investors and managers. Cultural tourism and cultural travel have changed from a luxury good to a consumer good used by the general population. When considering the establishment of Revived Originals and their many requirements, it is always necessary to compare these problems to those faced by the original cultural sites today and in the likely future.

Part III

Cultural Overtourism: A Perspective

8

Conclusions

8.1 Cultural Overtourism Is Controversial

Overtourism has been recognized and discussed as a problem in both traditional and new media. The severe negative effects have become obvious to many people: overuse and the resulting destruction of the cultural sites; emissions, pollution, waste, and noise; improper behavior and crime; and exploding traffic. In contrast, the benefits have been largely neglected. Without serious efforts to counteract overtourism, the impact on those affected by the sheer numbers will intensify.

Cultural tourists suffer from various negative consequences of overtourism. They consist of stress and queues, the partial destruction of cultural sites, ecological costs, and increased crime. At cultural sites, they mainly or even exclusively meet other tourists. The special ambience, which is much sought by many tourists interested in art, is lost, and the enjoyment of experiencing culture is reduced.

Local residents are also strongly affected and must leave the city centers, not least because rents are rising significantly and living conditions are becoming increasingly difficult due to mass tourism.

These circumstances have led to protests, particularly in Barcelona, Amsterdam, and Venice. But elsewhere too, opposition is growing among local populations, such as in Zermatt and Lucerne and on the Rigi and the Jungfraujoch in Switzerland. In 2018, 912,000 people visited the Rigi, and almost half a million the Jungfraujoch (Kamm-Sager 2019). Tourism organizations that profit from visitors are, of course, opposed to these protests. The winners benefiting from overtourism have a great deal of political clout, which gives them the ability to fight any restrictions on the number of tourists successfully.

The measures taken by governments to counter the negative effects of cultural overtourism have not been successful so far. Appeals and marketing efforts are not very effective. Therefore, they have to focus on administrative restrictions such as limiting access by time and place.

Increasingly, entrance fees are charged for visits to cultural sites. In February 2019, for example, Venice planned to charge 3 euros for each visitor; however, the start of the implementation has been repeatedly postponed. From 2020, it was planned to increase the entrance fees to 10 euros, depending on the season and the number of visitors intending to enter the town (Dignös 2019). In the period of undertourism during spring and summer 2020, the implementation was further postponed. Moreover, experience shows that imposing entrance fees has little effect because, especially for foreign tourists, the higher costs of a visit are negligible compared to the money spent on transport and hotels. Only if admission prices reach a high level does demand decrease. However, this measure is controversial because the procedure is not fair to lower-income groups. This could lead to the conclusion that attractive cultural locations are only affordable to the rich. It is also morally ambivalent whether future generations and travelers from other countries and continents may be, at least partially, excluded from enjoying art. Restricting demand by government interventions also has the undesirable effect of creating a black market for entry tickets.

To date, the measures taken to deal with overtourism have focused exclusively on demand. The number of tourists is to be reduced so that a visit can be more enjoyable for the others. One exception is the attempt to redirect tourists to cultural sites that have been rarely visited so far. But these alternative locations will also soon be overcrowded if the growth in tourism continues.

8.2   Cultural Overtourism: A Solution

My proposal of Revived Originals goes one significant step beyond current strategies. The supply of cultural sites needs to be increased by partially or completely rebuilding them nearly identically at suitable locations, so that the transport of visitors causes as little environmental damage as possible. The buildings reproduced by digital technology such as augmented and virtual reality must be linked to the historical and artistic context of the cultural site. Visitors should feel as if they were present at the time when the cultural sites were created. The culture exhibited must be explained in an easily understandable way. Historical and cultural experience conveyed in this way provides visitors to Revived Originals with an additional benefit that is sadly missing when visiting overcrowded original sites. The period of undertourism imposed by government policy to fight the COVID-19 pandemic has had one positive effect: otherwise badly overcrowded cities such as Florence, Rome, Paris, and especially Venice have appeared in a more appealing light when their squares and streets are deserted. Although the costs to local business are enormous, this experience demonstrates that reducing the number of tourists visiting the original sites is beneficial and should be taken into account when planning Revived Originals.

For the whole tourism industry, the idea of creating and operating Revived Originals opens up additional opportunities. Companies that offer cultural tours to tourists are no longer dependent on the original sites, which are often wrongly regarded as unique. If, for example, the public administration of Venice limits access to cultural sites (e.g., by imposing higher taxes or reducing admission times) for various reasons, an alternative in the form of a Revived Original of Venice situated nearby would make these restrictions superfluous. The travel organization is no longer dependent on a monopolistic provider and thus has more options to organize activities for culturally interested tourists. The existence of two cultural sites also creates additional jobs, which is beneficial for the tourism industry.

8.2.1 Revived Originals Will Cause Opposition

Revived Originals will meet substantial resistance from the existing local tourism industry. It will probably oppose such a proposal. It will fear a decline in the income and profits of hotels, restaurants, and shops located at the original site. This includes apartment owners who offer their flats on Airbnb.

Compared with established tourist tours, the historically replicated Revived Originals offer new options. It is not sure whether these alternative places will become reality at all; for the time being, it is only science fiction. Therefore, their supporters have less political influence than today's tourist resorts. This will continue if, as is to be expected, the total expenditure of tourists increases, because this will be divided between the two options of original sites and Revived Originals.

Some educated art lovers will view Revived Originals as a true sacrilege that must be fought vigorously. They will not accept this alternative, even if it is adeptly combined with history and culture. They will consider a strict distinction between original and copy to be crucial, even though this extreme differentiation has been abandoned in art history. A culturally important building has typically been altered over time. This development is normal for artworks and does not diminish their significance. This even applies to paintings that have been repainted in various ways.

Anyone who finds the Revived Originals inappropriate or even scandalous still has the opportunity to visit the original cultural site. The stay will even be more pleasant because some of the tourists will not go there anymore but visit the Revived Originals instead. The number of tourists visiting the original sites tends to decline. So educated citizens who still want to visit the originals should welcome the additional supply represented by Revived Originals.

Creating Revived Originals requires great effort. In addition to suitable locations concerning transport and the environment, it is also important to carefully select the buildings to be copied. Ideally, they should be arranged as a whole so that they stand in a context with each other. It is also difficult to combine a place of interest with local traditions, history, and culture. The latter is particularly important because it provides an additional benefit for culturally interested tourists. Only if this is implemented successfully will visitors no longer overcrowd the original places.

8.2.2   Revived Originals Offer Many Opportunities

It is important to present Revived Originals to visitors in an attractive way. For example, it would be wrong to prevent educated tourists from visiting the original site by diverting them to the replica Venice. Rather, the advantages of the Revived Originals must be highlighted. In particular, it should be emphasized that visitors to a Revived Original can immerse themselves in the history and art of the site. In addition, modern digital media used in the process will convey this unique selling point in tempting and surprising new ways. Moreover, tourists can enjoy these benefits without the stress caused by over-tourism through queues, crowding, pollution, and noise.

Tour operators need to consider which groups of tourists will benefit more from the Revived Originals. Obviously, these include young families, as the new offer focuses in particular on children and their interests. The same is true for tourists who would like to visit famous cultural sites but are not familiar with their art and history. This is probably true for a considerable part of tourists. In the Revived Originals, they are not lectured patronizingly, but outstanding experts explain the history and art of a cultural site in an easily understandable way. Another group who will possibly enjoy visiting the Revived Originals are young people. The wide range of digital technologies adopted, such as augmented reality, virtual reality, digital twins, and the depiction of famous people from the past by using holograms will particularly fascinate them. These advantages also apply to schools. Students normally find field trips to the original locations rather boring because their specific interests are not taken into account. Revived Originals focus exactly on this deficit. Lastly, Revived Originals give disabled and elderly people their first opportunity to comprehensively experience cities such as Dubrovnik or the Mont Saint Michel. These possibilities are limited today. Venice, for example, is scarcely accessible for handicapped people in wheelchairs because of the many bridges, which have to be climbed up and down again.

Tourism organizations can make the Revived Originals more attractive to visitors by using them as locations for films and TV series. In this way, a Revived Original will receive more attention and will be unforgettable to potential visitors. Such an effect can be expected especially when a film becomes cult. This has already been the case several times (see the comprehensive overview in Connell 2012).

- The television series *Game of Thrones* (2011–2019, with 73 episodes) was partly shot in Dubrovnik and is actively promoted in this city for marketing purposes. A carefully conducted econometric study (Depken et al. 2017) found a remarkable effect of this series on the number of visitors to the city: nearly 60,000 additional people each year have stayed overnight in Dubrovnik since *Game of Thrones* was aired.
- The film series *Sissi* of the 1950s, starring Romy Schneider, still attracts tourists to the places related to Empress Elisabeth. The Hofburg Palace in Vienna is widely marketed with Sissi's name today (Peters et al. 2011).
- The city of Vienna is closely associated with the film *The Third Man* (1949, directed by Carol Reed and written by Graham Greene), especially among Anglo-Saxon tourists. Many visitors very well remember the journey of the leading actor, Orson Welles, in a jeep through post-war Vienna.
- The Hollywood musical *The Sound of Music* (1965), filmed in and around Salzburg, still attracts many tourists from America because the film is frequently shown on television (Luger and East 2001).
- Venice is certainly also an attraction to fans of the film *Death in Venice*. The film was shot in 1971, based on the novella by Thomas Mann and directed by Luchino Visconti, with Dirk Bogarde playing the protagonist von Aschenbach. The film shows St. Mark's Square, the backyards abandoned due to cholera, and Aschenbach's crossing to the Lido by gondola.

Revived Originals, therefore, force providers of tourist tours to reinvent their business and to react more consciously in the interests of travelers. It is not enough to simply propose the Revived Originals or the traditional cultural sites. The different benefits of the two offers must be carefully presented and explained.

Revived Originals are no cure for all the problems associated with cultural overtourism. The idea and concept cannot be realized everywhere and may meet great difficulties. Like every concept, it also has its disadvantages. However, compared to today's obvious and often very disturbing situation in overcrowded cultural sites, Revived Originals are a promising step forward.

References

Abbasov, F. (2019). *One corporation to pollute them all*. Retrieved 19 August, 2019, from https://www.transportenvironment.org/publications/one-corporation-pollute-them-all.

Aichner, T., Maurer, O., Nippa, M., & Tonezzani, S. (2019). *Virtual reality im tourismus*. Wiesbaden: Springer Gabler.

Amore, A. (2019). *Tourism and urban regeneration: Processes compressed in time and space*. Abingdon, UK: Routledge.

Armellini, A. (2016). Touristenboom und Einwohnerschwund. *Venedig wird eine Geisterstadt*. Spiegel online. Retrieved from https://www.spiegel.de/reise/staedte/venedig-einwohner-demonstrieren-gegen-massentourismus-a-1121068.html.

Arnold, D. (2008). Digital artefacts: Possibilities and purpose. In M. Greengrass & L. Hughes (Eds.), *The virtual representation of the past* (pp. 160–170). London: Routledge.

Aronson, A. (2018). *The Routledge companion to scenography*. London: Routledge.

Ashworth, G., & Page, S. J. (2011). Urban tourism research: Recent progress and current paradoxes. *Tourism Management, 32*(1), 1–15. https://doi.org/10.1016/j.tourman.2010.02.002.

Association of Mediterranean Cruise Ports. (2018). Retrieved from https://www.medcruise.com/18-statistics-cruise-activities-in-medcruise-ports.

Barron, K., Kung, E., & Proserpio, D. (2018). The sharing economy and housing affordability: Evidence from airbnb. *SSRN Electronic Journal*. https://doi.org/10.2139/ssrn.3006832.

Basilica di San Marco. (2020). *The building phases*. Retrieved 02 September, 2020, from http://www.basilicasanmarco.it/basilica/architettura/fasi-costruttive/?lang=en.

© The Author(s), under exclusive license to Springer Nature Switzerland AG 2021
B. S. Frey, *Overcoming Overtourism*, https://doi.org/10.1007/978-3-030-63814-6

Baudrillard, J. (1994). *Simulacra and simulation*. Ann Arbor: University of Michigan Press.

Becker, E. (2013). *Overbooked: The exploding business of travel and tourism*. New York: Simon & Schuster.

Belisle, F. J., & Hoy, D. R. (1980). The perceived impact of tourism by residents: A case study in Santa Marta, Colombia. *Annals of Tourism Research, 7*(1), 83–101. https://doi.org/10.1016/S0160-7383(80)80008-9.

Bellon, T. (2018). Berlin loosens law for short-term home rentals. *Reuters*. Retrieved 22 August, 2019, from https://www.reuters.com/article/airbnb-berlin/berlin-loosens-law-for-short-term-home-rentals-idUSL8N1R473J.

Benz, M. (2019). Hallstatt–Das überfüllte "Paradies". *Neue Zürcher Zeitung*, 23 August: 26–27.

Brida, J. G., & Zapata, S. (2010). Cruise tourism–economic, sociocultural and environmental impacts. *International Journal of Leisure and Tourism Marketing, 1*(3), 205–226.

Bundesanstalt Statistik Österreich. (2019). *Nächtigungsstatistik ab 1974 (GEH) Q, [tou_int_geh]*.

Butler, D. (2008). Architecture: architects of a low-energy future. *Nature, 452*(7187), 520–524. https://doi.org/10.1038/452520a.

Città di Venezia. (2019, 17 October). *#EnjoyRespectVenezia*. Retrieved 10 April, 2020, from https://www.comune.venezia.it/de/content/enjoyrespectvenezia.

Città di Venezia. (2020, 3 January). *Touristeninformation–Hochrechnung der pro Tag in Venedig erwarteten Besucher*. Retrieved 10 April, 2020, from https://www.comune.venezia.it/de/content/touristeninformation-hochrechnung-der-pro-tag-venedig-erwarteten-besucher.

Clancy, M. (2019). Chapter 2: Overtourism and resistance. Today's anti-tourist movement in context. In H. Pechlaner, E. Innerhofer, & G. Eschbaner (Eds.), *Overtourism, tourism management and solutions*. London/New York: Routledge.

Clemente-Ruiz, A., & Aloudat, N. (Eds.). (2019). *Von Mossul nach Palmyra. Eine Virtuelle Reise durch das Weltkulturerbe*. München: Hirmer.

Connell, J. (2012). Film tourism–evolution, progress and prospects. *Tourism Management, 33*(5), 1007–1029. https://doi.org/10.1016/j.tourman.2012.02.008.

Connolly, K. (2019). A rising tide: "Overtourism" and the curse of the cruise ships. *Guardian*. Retrieved 16 September, 2019, from https://www.theguardian.com/business/2019/sep/16/a-rising-tide-overtourism-and-the-curse-of-the-cruise-ships.

Croce, V. (2018). With growth comes accountability: Could a leisure activity turn into a driver for sustainable growth? *Journal of Tourism Futures, 4*(3), 218–232. https://doi.org/10.1108/JTF-04-2018-0020.

D'Eramo, M. (2017). *Il Selfie del Mondo. Indagine sull'Età del Turismo*. Milano: Feltrinelli.

D'Eramo, M. (2018). *Der Reisende ist nur ein Tourist, der abstreitet, einer zu sein. Interview by Daniel Weber, NZZFolio, Wir Touristen, October*. Retrieved from https://folio.nzz.ch/2018/oktober/der-reisende-ist-nur-ein-tourist-der-abstreitet-einer-zu-sein.

Daponte, P., De Vito, L., Picariello, F., & Ricio, M. (2014). State of the art and future developments of augmented reality for measurement applications. *Measurement, 57*, 53–70.

De Lusenet, Y. (2007). Tending the garden or harvesting the fields: Digital preservation and the UNESCO charter on the preservation of the digital heritage. *Library Trends, 56*(1), 164–182. https://doi.org/10.1353/lib.2007.0053.

Depken, C. A., Globan, T., & Kožić, I. (2017). Television Induced Tourism: Evidence from Croatia. *SSRN*. Retrieved from https://ssrn.com/abstract=3002690.

Dignös, E. (2019). Wie sich überlaufene Urlaubsorte wehren. *Süddeutsche Zeitung Magazin*. Reteieved 23 April, 2019, from https://www.sueddeutsche.de/reise/tourismus-massentourismus-strategien-gegen-overtourism-1.4407703.

Dodds, R., & Butler, R. (Eds.). (2019). *Overtourism: Issues, realities and solutions*. Berlin/Boston: De Gruyter.

Du Cros, H., & McKercher, B. (2020). *Cultural tourism* (3rd ed.). London: Routledge.

Duval, M., Smith, B. W., Gauchon, C., Mayer, L., & Malgat, C. (2019). "I have visited the Chauvet Cave": The heritage experience of a rock art replica. *International Journal of Heritage Studies, 26*, 142–162. https://doi.org/10.1080/13527258.2019.1620832.

Fawcett, J. (1998). Use and abuse: Management and good practice in cathedrals and greater churches. In J. Fawcett (Ed.), *Historic floors: Their history and conservation* (pp. 120–128). Oxford: Butterworth-Heinemann.

FAZ. (2019). *So viel Eintritt müssen Venedig-Touristen künftig zahlen*. Retrieved 14 February, 2019, from https://www.faz.net/aktuell/gesellschaft/menschen/italien-so-viel-eintritt-muessen-venedig-touristen-kuenftig-zahlen-16024718.html.

Flughafen Zürich. (2020). *Bewegungsstatistik*. Retrieved 4 April, 2020, from https://www.flughafen-zuerich.ch/unternehmen/laerm-politik-und-umwelt/flugbewegungen/bewegungsstatistik.

Frey, B. S. (2003). *Arts & economics–analysis & cultural policy* (2nd ed.). Berlin, Heidelberg: Springer.

Frey, B. S. (2019). *Economics of art and culture*. Cham: Springer.

Frey, B. S. (2020). Venedig ist überall. In *Vom Übertourismus zum Neuen Orginal*. Cham: Springer.

Frey, B. S., & Briviba, A. (2020). Revived originals—A proposal to deal with cultural overtourism. *Tourism Economics*. https://doi.org/10.1177/1354816620945407.

Frey, B. S., & Meier, S. (2006). The economics of museums. In V. A. Ginsburgh & D. Throsby (Eds.), *Handbook of the economics of art and culture* (Vol. 1, pp. 1017–1047). North Holland: Elsevier.

Frey, B. S., & Pamini, P. (2009). Making world heritage truly global: The culture certificate scheme. *Oxonomics, 4*(2), 1–9. https://doi.org/10.1111/j.1752-5209.2009.00033.x.

Frey, B. S., & Pommerehne, W. W. (1993). On the fairness of pricing–An empirical survey among the general population. *Journal of Economic Behavior and Organization, 20*(3), 295–307.

Frey, B. S., & Steiner, L. (2011). World heritage list: Does it make sense? *International Journal of Cultural Policy, 17*(5), 555–573. https://doi.org/10.1080/1028663 2.2010.541906.

García-Hernández, M., De la Calle-Vaquero, M., & Yubero, C. (2017). Cultural heritage and urban tourism: Historic city centres under pressure. *Sustainability, 9*(8), 1–19. https://doi.org/10.3390/su9081346.

García-Palomares, J., Gutiérrez, J., & Mínguez, C. (2015). Identification of tourist hot spots based on social networks: A comparative analysis of European metropolises using photo-sharing services and GIS. *Applied Geography, 63*, 408–417. https://doi.org/10.1016/j.apgeog.2015.08.002.

Giuffrida, A. (2019). *The death of Venice? City's battles with tourism and flooding reach crisis level.* Retrieved 11 August, 2019, from https://www.theguardian.com/world/2019/jan/06/venice-losing-fight-with-tourism-and-flooding.

Goodwin, H. (2017). The challenge of overtourism. *Responsible Tourism Partnership.* Working Paper 4. Retrieved from https://haroldgoodwin.info/pubs/RTP%27WP4Overtourism01%272017.pdf.

Gössling, S., Scott, D., & Michael Hall, C. (2018). Global trends in length of stay: Implications for destination management and climate change. *Journal of Sustainable Tourism, 26*(12), 2087–2101. https://doi.org/10.1080/0966958 2.2018.1529771.

Greengrass, M., & Hughes, L. (2008). *The virtual representation of the past.* London/New York: Routledge.

Hafner, U. (2019). *Venice time machine: Streit um Millionen-Projekt.* Retrieved from https://nzzas.nzz.ch/wissen/venice-time-machine-knatsch-um-millionen-projekt-ld.1528382?reduced=true.

Hardin, G. (1968). The tragedy of the commons. *Science, 162*(3859), 1243–1248.

Hawkins, D. E., Chang, B., & Warnes, K. (2009). A comparison of the national geographic stewardship scorecard ratings by experts and stakeholders for selected world heritage destinations. *Journal of Sustainable Tourism, 17*(1), 71–90. https://doi.org/10.1080/09669580802209944.

Heim, C. (2019). "Wovon man nicht sprechen kann, darüber soll man schweigen." *Tagesanzeiger. Das Magazin,* no. 42, 19 October: 24–27.

Hospers, G.-J. (2019). Overtourism in European cities: From challenges to coping strategies. *CESifo Forum, 20*(3), 20–24.

Howard, P. (2009). *What is scenography?* (2nd ed.). London: Routledge.

Hughes, N. (2018). "Tourists Go Home": Anti-tourism industry protest in Barcelona. *Social Movement Studies, 17*(4), 471–477. https://doi.org/10.1080/1474283 7.2018.1468244.

IUHB. (2019). *Touristik-Radar. Wie Urlauber auf Overtourismus reagieren.* Deutschland: Internationale Hochschule. Retrieved from https://www.iubh-university.de/wp-content/uploads/IUBH_Themenmappe-Overtourism.pdf.

Jacobsen, J. K. S., Iversen, N. M., & Hem, L. E. (2019). Hotspot crowding and overtourism: Antecedents of destination attractiveness. *Annals of Tourism Research, 76*, 53–66.

Jin, Q., Hu, H., & Kavan, P. (2016). Factors influencing perceived crowding of tourists and sustainable tourism destination management. *Sustainability, 8*(10), 976. https://doi.org/10.3390/su8100976.

Joost, M. (2019). *Chinas Tower-Bridge-Kopie sorgt für Unmut.* https://www.geo.de/reisen/reiseziele/15909-rtkl-suzhou-chinas-tower-bridge-kopie-sorgt-fuer-unmut.

Jung, J. (2020). *Das Laboratorium des Fortschritts. Die Schweiz im 19. Jahrhundert.* Basel: NZZ Libro.

Kahneman, D., Knetsch, J. L., & Thaler, R. (1986). Fairness as a constraint on profit seeking. Entitlements in the market. *American Economic Review, 76*(4), 728–741.

Kaminski, J., Benson, A. M., & Arnold, D. (2013). *Contemporary issues in cultural heritage tourism.* New York/Abingdon: Routledge.

Kamm-Sager, C. (2019). (Zu) viele Touristen: Diese Orte sind überfüllt, haben capituliert oder sind ganz geschlossen für die Massen. *St. Galler Tagblatt,* 8 October. https://www.tagblatt.ch/leben/zu-viele-touristen-diese-orte-sind-ueberfuellt-haben-kapituliert-oder-sind-ganz-geschlossen-fuer-die-massen-ld.1158124.

Kester, J. (2016). International Tourism Trends in EU-28 Member States: Current Situation and Forecasts for 2020-2025-2030. Retrieved 12 August, 2019, from http://www.eufed.org/binary/uploads//UNWTO_TT2030_EU28.pdf.

Koens, K., Postma, A., & Papp, B. (2018). Is overtourism overused? Understanding the impact of tourism in a city context. *Sustainability, 10*(12), 4384. https://doi.org/10.3390/su10124384.

Landorf, C. (2009). Managing for sustainable tourism: A review of six cultural world heritage sites. *Journal of Sustainable Tourism, 17*(1), 53–70.

Larson, L. R., & Poudyal, N. C. (2012). Developing sustainable tourism through adaptive resource management: A case study of Machu Picchu, Peru. *Journal of Sustainable Tourism, 20*(7), 917–938. https://doi.org/10.1080/0966958 2.2012.667217.

Leslie, D., & Sigala, M. (2005). *International cultural tourism.* Oxford: Butterworth-Heinemann.

Losse, B. (2019). Man sollte Venedig als Replik nachbauen. *Wirtschafts Woche/Der Volkswirt, 25*, 40–41.

Luger, K., & East, P. (2001). Living in paradise: Youth culture and tourism development in the mountains of Austria. In R. Voase (Ed.), *Tourism in Western Europe—A collection of case histories* (pp. 227–242). Wallingford, UK/New York, NY: CABI Publishing.

Lundberg, D. E. (1990). *The tourist business* (6th ed.). New York: Van Nostrand Reinhold.

Marti, G. A. (2019). Komodoinsel wird zum teuren Pflaster. Hoher Eintrittspreis soll Touristenstrom eindämmen. *Neue Zürcher Zeitung,* Donnerstag, 3 October: 24. https://www.nzz.ch/panorama/indonesien-will-dracheninsel-komodo-teurer-machen-ld.1512620?reduced=true.

Martín, M., María, J., Martínez, J. M. G., & Fernández, J. A. S. (2018). An analysis of the factors behind the Citizen's attitude of rejection towards tourism in a context of overtourism and economic dependence on this activity. *Sustainability, 10*(8), 2851. https://doi.org/10.3390/su10082851.

Martinez, J.-L. (2019). Welche Zukunft haben die zerstörten Monumente? In A. Clemente-Ruiz & N. Aloudat (Eds.), *Von Mossul nach Palmyra. Eine virtuelle Reise durch das Weltkulturerbe* (pp. 112–117). München: Hirmer.

Martín-Martín, J. M., Ostos-Rey, M. S., & Salinas-Fernández, J. A. (2019). Why regulation is needed in emerging markets in the tourism sector. *American Journal of Economics and Sociology, 78*(1), 225–254. https://doi.org/10.1111/ajes.12263.

McKinney, J., & Butterworth, P. (2009). *The Cambridge introduction to scenography.* Cambridge: Cambridge University Press.

McKinsey & Company and World Travel & Tourism Council. (2017). *Coping with success.* London: Managing Overcrowding in Tourism Destinations. Retrieved from https://www.mckinsey.com/industries/travel-transport-and-logistics/our-insights/coping-with-success-managing-overcrowding-in-tourism-destinations.

Metz, M., & Seesslen, G. (2019, 19 May). *Sonderangebote—"Kreative" Preisgestaltung soll Kauf-Impulse auslösen.* Retrieved 10 April, 2020, from https://www.deutsch-landfunk.de/sonderangebote-kreative-preisgestaltung-soll-kauf-impulse.1184.de.html?dram:article_id=444842.

Milano, C. (2017). Overtourism and tourismphobia: Global trends and local contexts. In *Technical Report.* Barcelona: Ostelea School of Tourism & Hospitality.

Milano, C. (2018). Overtourism, Malestar Social y Turismofobia. Un Debate Controvertido. *PASOS. Revista de Turismo y Patrimonio Cultural, 16*(3), 551–564.

Milano, C., Cheer, J. M., & Novelli, M. (2018). *Overtourism: A growing global problem.* Retrieved 12 August, 2019, from https://theconversation.com/overtourism-a-growing-global-problem-100029.

Milano, C., Novelli, M., & Cheer, J. M. (2019a). Overtourism and degrowth: A social movements perspective. *Journal of Sustainable Tourism, 27*(12), 1857–1875. https://doi.org/10.1080/09669582.2019.1650054.

Milano, C., Cheer, J. M., & Novelli, M. (Eds.). (2019b). *Overtourism: Excesses, discontents and measures in travel and tourism.* Wallingford, Oxfordshire, Boston, MA: CABI.

Muscheler, U. (2008). *Die Nutzlosigkeit des Eiffelturms: Eine etwas andere Architekturgeschichte.* München: Beck.

Neuhaus, C. (2019). Interview. Tourismusexperte Christian Laesser: "Mittlerweile gibt es auch in der Schweiz Orte, wo die Grenze zum Overtourismus erreicht ist." *Neue Zürcher Zeitung.* Retrieved 15 May, 2019, from https://www.nzz.ch/schweiz/movertourism-es-gibt-schweizer-orte-wo-die-grenze-erreicht-ist-nzz-ld.1482167?reduced=true.

OECD/ICOM. (2018). *Culture and local development: Maximising impact. Guide for local governments, communities and museums. Launch Version.* Paris: OECD and ICOM.

OECD/ICOM. (2019). *Culture and local development: Maximizing the impact. A guide for local governments, communities and museums. 2018 OECD Conference on Culture and Local Development.* Paris: OECD. Retrieved from https://www.oecd-ilibrary.org/docserver/9a855be5-en.pdf?expires=1587996291&id=id&accname=guest&checksum=6774A967D4A9C5A7BCA470E71C03B630.

Ostrom, E. (1990). *Governing the commons: The evolution of institutions for collective action.* Cambridge: Cambridge University Press.

Pan, W. (2014). System boundaries of zero carbon buildings. *Renewable and Sustainable Energy Reviews, 37*, 424–434. https://doi.org/10.1016/j.rser.2014.05.015.

Pechlaner, H., Innerhofer, E., & Eschbaner, G. (Eds.). (2019). *Overtourism. Tourism management and solutions.* London and New York: Routledge.

Pedersen, I., Gale, N., Mirza-Babaei, P., & Reid, S. (2017). More than meets the eye: The benefits of augmented reality and holographic displays for digital cultural heritage. *Journal on Computing and Cultural Heritage, 10*(2), 1–15.

Peeters, P., Gössling, S., Klijs, J., Milano, C., Novelli, M., Dijkmans, C., Eijgelaar, E., Hartman, S., Heslinga, J., Isaac, R., Mitas, O., Moretti, S., Nawijn, J., Papp, B., & Postma, A. (2018). *Research for TRAN committee-overtourism: Impact and possible policy responses.* Brussels: European Parliament, Policy Department for Structural and Cohesion Policies.

Peters, M., Schuckert, M., Chon, K., & Schatzmann, C. (2011). Empire and romance: Movie-induced tourism and the case of the sissi movies. *Tourism Recreation Research, 36*(2), 169–180. https://doi.org/10.1080/0250828 1.2011.11081317.

Phelan, J. (2018, 30 August). *In just nine months, nearly 500 florence residents were turfed out to make way for tourist rentals.* Retrieved 6 April, 2020, from https://www.thelocal.it/20180830/airbnb-florence-tourist-rental-evictions.

Picascia, S., Romano, A., & Teobaldi, M (2017). The airification of cities: Making sense of the impact of peer to peer short term letting on urban functions and economy. *Proceedings of the annual congress of the association of european schools of planning, Lisbon 11–14 July.*

Pietersen, P. (2006). *Kriegsverbrechen der alliierten Siegermächte: Terroristische Bombenangriffe auf Deutschland und Europa 1939–1945.* Norderstedt: BoD Books on Demand.

Pitrelli, M. B. (2019, December 5). *Instagrammers love this iconic spot, but there's something they don't want you to see.* Retrieved 5 April, 2020, from https://www.cnbc.com/2019/12/02/norways-social-media-hot-spots-trolltunga-preikestolen-and-kjeragbolten.html.

Pousset, S. (2019). Herdentrieb–Wer rettet das Reisen vor Instagram? *Frankfurter Allgemeine Quarterly, 4*, 18.

Ram, Y., & Michael Hall, C. (2017). *Walkable places for visitors: Assessing and designing for walkability. The Routledge International Handbook of Walking* (pp. 311–329). London/New York: Routledge.

Rasoolimanesh, S. M., Roldán, J. L., Jaafar, M., & Ramayah, T. (2017). Factors influencing residents' perceptions toward tourism development: Differences across rural and urban world heritage sites. *Journal of Travel Research, 56*(6), 760–775. https://doi.org/10.1177/0047287516662354.

Reinhardt, V. (2019). Das Selfie verheisst Unsterblichkeit. *Neue Zürcher Zeitung, 18*, 37.

Reski, P. (2013). *Kulturkampf–Die verkaufte Seele der Lagune. Tagespiegel.* Retrieved 5 April, 2020, from https://www.tagesspiegel.de/kultur/kulturkampf-die-verkaufte-seele-derlagune/8026522.html.

Responsible Travel and Google. (2019). *Overtourism mapped: Tourism is headed into a global crisis.* Retrieved 28 October, 2020, from https://www.responsibletravel.com/copy/overtourism-map.

Richards, G. (1996). *Cultural tourism in Europe.* Welllingford, UK: CAB International.

Richards, G. (2014). *Tourism trends: The convergence of culture and tourism.* The Netherlands: Academy for Leisure NHTV University of Applied Sciences.

Richards, G. (2018). Cultural tourism: A review of recent research and trends. *Journal of Hospitality and Tourism Management, 36*, 12–21. https://doi.org/10.1016/j.jhtm.2018.03.005.

Richardson, D. (2017). *WTM 2017: Europe suffering the strain of tourism.* Retrieved from https://www.ttgmedia.com/wtm-news/wtm-news/wtm-2017-europe-suffering-the-strain-of-tourism-12206.

Rosen, S. (1981). The economics of superstars. *American Economic Review, 71*(5), 845–858.

Sans, A. A., & Quaglieri, A. (2016). Unravelling airbnb: Urban perspectives from Barcelona. In A. P. Russo & G. Richards (Eds.), *Reinventing the local in tourism: Producing, consuming and negotiating place* (pp. 209–228). Bristol: Channel View Publications.

Schader, A., & Aeby, N. (2019). Wie es Gott gefällt: Streifzug durchs christliche Amerika—ein Fototableau von Cyril Abad. *Neue Zürcher Zeitung,* Retrieved 14 October, 2019, from https://www.nzz.ch/feuilleton/wie-es-gott-gefaellt-streifzug-durchs-christliche-amerika-ein-foto-tableau-von-ld.1514378?reduced=true.

Senn, L., & Egger, M. (2019). *Diese 5 Grafiken zeigen, wie rasant der Tourismus angestiegen ist.* Retrieved from https://www.watson.de/leben/reisen/296819127-diese-5-grafiken-zeigen-wie-rasant-der-tourismus-angestiegen-ist.

Seraphin, H., Yallop, A. C., Capatîna, A., & Gowreesunkar, V. (2018a). Heritage in tourism organisations' branding strategy: The case of a post-colonial, post-conflict and post-disaster destination. *International Journal of Culture Tourism and Hospitality Research, 12*(1), 89–105. https://doi.org/10.1108/IJCTHR-05-2017-0057.

Seraphin, H., Sheeran, P., & Manuela Pilato, M. (2018b). Over-tourism and the fall of venice as a destination. *Journal of Destination Marketing & Management, 9,* 374–376. https://doi.org/10.1016/j.jdmm.2018.01.011.

Seraphin, H., Zaman, M., Olver, S., Bourliataux-Lajoinie, S., & Dosquet, F. (2019). Destination branding and overtourism. *Journal of Hospitality and Tourism Management, 38,* 1–4. https://doi.org/10.1016/j.jhtm.2018.11.003.

Singh, T. (2018). Is over-tourism the downside of mass tourism? *Tourism Recreation Research, 43*(4), 415–416. https://doi.org/10.1080/02508281.2018.1513890.

Smeral, E. (2019). Chapter 12: Overcrowding in tourism destinations. In H. Pechlaner, E. Innerhofer, & G. Eschbaner (Eds.), *Overtourism. Tourism management and solutions.* London/New York: Routledge.

Smith, M., & Richards, G. (2013). *The Routledge handbook in cultural tourism.* London: Routledge.

Steiger, R., Scott, D., Abegg, B., Pons, M., & Aall, C. (2019). A critical review of climate change risk for ski tourism. *Current Issues in Tourism, 22*(11), 1343–1379. https://doi.org/10.1080/13683500.2017.1410110.

Steinecke, A. (2010). Culture—A tourist attraction: Importance–expectations–potential. In R. Conrady & M. Buck (Eds.), *Trends and issues in global tourism 2010* (pp. 185–196). Berlin, Heidelberg: Springer.

Su, Y.-W., & Lin, H.-L. (2014). Analysis of international tourist arrivals worldwide: The role of world heritage sites. *Tourism Management, 40,* 46–58. https://doi.org/10.1016/j.tourman.2013.04.005.

Thani, S., & Heenan, T. (2017). The ball may be round but football is becoming increasingly arabic: Oil money and the rise of the new football order. *Soccer & Society, 18*(7), 1012–1026.

The Art Newspaper. (2019). *Special report, Number 311, April,* https://www.museus.gov.br/wp-content/uploads/2019/04/The-Art-Newspaper-Ranking-2018.pdf.

The Economist. (2019). *Social media. Daka destinations. 17 August: 42.* https://www.economist.com/china/2019/08/15/for-some-in-china-the-aim-of-travel-is-to-create-15-second-videos.

The Economist. (2020a). *Traffic on everest. High and climbing. 14 March: 40.* https://www.economist.com/asia/2020/03/12/new-rules-to-limit-numbers-on-everest-are-delayed.

The Economist. (2020b). *The data economy: Mirror worlds. A deluge of data is giving rise to a new economy. 22 February: 3–4.* Retrieved from https://www.economist.com/special-report/2020/02/20/a-deluge-of-data-is-giving-rise-to-a-new-economy.

Throsby, D. (2001). *Economics and culture.* Cambridge: Cambridge University Press.

Tosun, C. (2002). Host perceptions of impacts: A comparative tourism study. *Annals of Tourism Research, 29*(1), 231–253. https://doi.org/10.1016/S0160-7383(01)00039-1.

UNESCO. (2019). *Operational guidelines for the implementation of the world heritage convention.* Retrieved from https://whc.unesco.org/en/guidelines.

UNWTO. (2018). *Overtourism? Understanding and managing urban tourism growth beyond perceptions.* Madrid: UNWTO. https://doi.org/10.18111/9789284420070.

Vecco, M., & Caus, J. (2019). Chapter 5: UNESCO, cultural heritage sites and tourism. A paradoxical relationship. In H. Pechlaner, E. Innerhofer, & G. Eschbaner (Eds.), *Overtourism. Tourism management and solutions.* London/New York: Routledge.

Vianello, M. (2016). The no grandi navi campaign: Protests against cruise tourism in venice. In C. Colomb & J. Novy (Eds.), *Protest and resistance in the tourist city* (pp. 171–190). London: Routledge.

Weaver, D., & Lawton, L. J. (2001). Resident perceptions in the urban–rural fringe. *Annals of Tourism Research, 28*(2), 439–458. https://doi.org/10.1016/S0160-7383(00)00052-9.

Weber, F. (2017). Overtourism. An Analysis of Contextual Factors Contributing to Negative Developments in Overcrowded Tourism Destinations. *BEST EN Think Tank XVII: Innovation and progress in sustainable tourism: Conference Proceedings,* Mauritius. pp. 315–320.

Weber, F., Stettler, J., Priskin, J., Rosenberg-Taufer, B., Ponnapureddy, S., Fux, S., Camp, M.-A., & Barth, M. (2017). *Tourism destinations under pressure. Challenges and innovative solutions.* Switzerland: Lucerne University of Applied Sciences and Arts. Retrieved from https://static1.squarespace.com/static/56dacbc6d210b821510cf939/t/5906f320f7e0ab75891c6e65/1493627704590/WTFL_study+2017_full+version.pdf.

Wong, L., & Quintero, M. S. (2019). Tutankhamen's two tombs: Replica creation and the preservation of our cultural heritage in the digital age. *The International Archives of the Photogrammetry, Remote Sensing and Spatial Information Sciences, 42-2/W11*, 1145–1150.

Zeng, B., & Gerritsen, R. (2014). What do we know about social media in tourism? A review. *Tourism Management Perspectives, 10*, 27–36. https://doi.org/10.1016/j.tmp.2014.01.001.

Zerva, K., Palou, S., Blasco, D., & Donaire, J. A. B. (2019). Tourism-philia versus tourism-phobia: Residents and destination management organization's publicly expressed tourism perceptions in Barcelona. *Tourism Geographies, 21*(2), 306–329. https://doi.org/10.1080/14616688.2018.1522510.